One Step Sideways,
Three Steps Forward

Crossing an *a'a* lava field on Isabela Island with Peter (photo by Martin Wikelski).

One Step Sideways, Three Steps Forward

ONE WOMAN'S PATH TO
BECOMING A BIOLOGIST

B. ROSEMARY GRANT

PRINCETON UNIVERSITY PRESS
PRINCETON & OXFORD

Published by Princeton University Press
41 William Street, Princeton, New Jersey 08540
99 Banbury Road, Oxford OX2 6JX

press.princeton.edu

ISBN 978-0-691-26059-4
ISBN (e-book) 978-0-691-26069-3

British Library Cataloging-in-Publication Data is available

Editorial: Alison Kalett and Hallie Schaeffer
Production Editorial: Jill Harris
Jacket Design: Heather Hansen
Production: Danielle Amatucci
Publicity: Matthew Taylor and Kate Farquhar-Thomson
Copyeditor: Amy K. Hughes

Jacket photograph by Roger Berglund

This book has been composed in Arno & Sans

Printed on acid-free paper. ∞

Printed in the United States of America

10 9 8 7 6 5 4 3 2 1

For Peter

Our daughters, Nicola and Thalia

And grandchildren, Rajul, Orange, Olivia, and Devon

CONTENTS

PREFACE

I joined the world's two billion human inhabitants on October 8, 1936. By November 2022, the human population had exceeded eight billion. Between these two times, when our numbers multiplied fourfold, our exceptional ability to communicate with each other across the globe resulted in outstanding discoveries and innovations in science, medicine, and the arts. But at the same time we have witnessed wars, habitat destruction, loss of biodiversity, global pollution, and increased inequalities between peoples. This book is a retrospective view of how my life unfolded against this rich tapestry of change.

Experiences in early life shape our interests and govern our actions in later life. I was born in Arnside, Westmorland, a small, sleepy, coastal village of a few hundred inhabitants in the mountainous Lake District of North West England. Famous for its tidal bore, which roars up the estuary twice a day, Arnside has fossil-rich carboniferous limestone cliffs backed by wooded valleys and high cropped-grass fells on which rare species of plants are pollinated by butterflies and bees. I roamed these woods and fells with my mother, learning the names of plants and habits of birds, and at a childish level I wondered how such a rich diversity of life came to be. Finding fossils in the Arnside rocks that were similar and yet differed in many features from their present-day living relatives added a deeper perspective to this question. Life had not been static over time—how and why had it changed? These evolutionary questions arose in embryonic form in my early childhood.

This fascination with evolution, originating in the rich natural history of Arnside, was fostered by my parents' questions; by the insights of a gardener, Jeremiah Swindlehurst; and by my reading of Charles Darwin's *Origin of Species* at a young age. The explicit thread running through my life is an interest in variation within populations of animals and plants—that is, the numerous differences between individuals in the same population, including us humans. How does this variation in physical and social environments give rise to change? How and why do species multiply? Such questions steered me toward studying genetics, ecology, behavior, and evolution at university, and finally to exploring the processes involved in the formation of new species. I accomplished the last goal in scientific research in the Galápagos with my husband, Peter.

At another level, questions of variation gave rise to an interest in the origin of the human condition: how and why humans moved across the globe and how groups with different backgrounds have mingled and exchanged genes and cultural influences. Why, I wondered, have these meetings of ideas sometimes led to a tolerance of differences, cooperation, and an explosion of creativity, whereas on other occasions they resulted in competition, racism, inequities between peoples, abuse of the environment, and war. I have been preoccupied with the challenge of understanding the causes and solutions that could lead to cultivating the former while avoiding the latter. Here, I believe, education has a significant role to play.

My path was neither conventional nor straight. I was born in an era when few women worked professionally, and there were many times when my goal of becoming a biologist was thwarted. Sidestepping and circumnavigating obstructions often took me on a different trajectory that turned out unexpectedly to be superior to the one intended. Eventually, I was able to explore deeply those early childhood questions.

I hope my journey will inspire others to follow their dreams, people of all ages, from all backgrounds and circumstances, including those like me who wish to juggle parenting with professional life and have the joy of working closely with a partner, sharing ideas, disappointments, and successes. Whatever their chosen interest, at whatever time in their life, their journey, like mine, will not be smooth, but if they follow their aspirations and heart, there will be abundant rewards and a touch of magic. It is for them that I have written this book.

PART I

Childhood

1

Early Years

> There is grandeur in this view of life, . . . whilst this planet has gone cycling on according to the fixed law of gravity, from so simple a beginning endless forms most beautiful and most wonderful have been, and are being, evolved.
>
> —CHARLES DARWIN, *ON THE ORIGIN OF SPECIES*

I entered the Holocene with a bump. My earliest memories are of falling out of a high chair, having my head stitched, and seeing patients waiting to be attended to by my father. So began my happy years of childhood in the English Lake District.

I was named Barbara Rosemary Matchett. However, shortly after my birth certificate was signed, I became Rosemary, my mother having discovered that Barbara was the name of my father's previous girlfriend. I still feel a tinge of her indignation at airports and doctors' offices when I am referred to as Barbara.

My parents were central figures in my early life. My mother was vivacious, energetic, and warmly loving. She introduced me to the joys of listening to the complexities of classical music, and on our daily walks she told me the names and stories of fossils, wild plants, and birds and other animals. My father, a country doctor, had his "surgery" and office in a wing of the house. Those were the days before disposable medical equipment, and much of my mother's time was spent sterilizing instruments and boiling surgical towels, as well as calming frightened patients and arranging appointments—all this while maintaining a large house and a garden with vegetables, fruit trees, hens, and pigs, and keeping control of two, and later three, very lively children.

My father was kind, but his demeanor could be severe. At school, he had received a classical education in Greek and Latin and had a broad knowledge of literature and poetry. He was trained in medicine in Glasgow at a time when tuberculosis was rampant. The disease was so serious and transmission so rapid that an extra year was added to the medical curriculum solely for teaching its diagnosis and treatment. He wanted to become a surgeon, and certainly this was within reach, as he was one of the top three students in his year, graduating *cum laude*. He opted instead for general practice (family medicine), knowing that it would give him a wide range of medical experience, and because he had to earn money quickly, as he was the second-youngest of his large family. Family medicine was a fortunate choice because it turned out he was unusually gifted in diagnosis. His exceptional skill in diagnosing complicated medical problems became widely known, and hospitals in London occasionally contacted him for help with a particularly difficult case. In addition, he had a passion for engineering. He designed and contributed to the first disposable syringe, now labeled as the "Matchett" syringe in the Science Museum of the Wellcome Collection in London, as well as other medical equipment.[1] His first practice was in partnership with a Dr. Patchett. We used to tease him that "Matchett and Patchett" would have made superb plastic surgeons.

My brother John, twenty-two months younger than me, and I were constantly in each other's company, there being few children in Arnside. The garden became our secret world, divided into countries with people in each speaking a different, invented language. We lived for hours in this world, going from cold Arctic to hot steamy jungle, creeping through the long orchard grass to avoid tigers, riding camels across the desert, climbing precipitous mountains, and vaulting over rushing mountain becks. Interestingly, our imaginary world with its diversity of countries had *no* wars, even though the Second World War was going on around us in the real world and impacting our lives in many ways (chapter 2).

That real world was the small village of Arnside, on the southern side of the Kent estuary in the southwest corner of England's Lake District. Arnside Knott, a hill only 522 feet high, rises steeply above the village. A climb up its slopes to the top reveals magnificent views across the estuary and beyond to Coniston Old Man, hunched in the foreground, and the other Lake District mountains behind it, each with its evocative Old English name. Throughout my childhood, horses were used to plow the fields and deliver milk and coal. Dr. Grosvenor, my father's predecessor, visited his patients on horseback. In the 1930s, the village had three telephones, the numbers being 1 for the pub, 2 for the vicar, and 3 for the doctor, my father. This shows the priorities of the inhabitants! More telephones were gradually added over time, and so was a telephone exchange as a hub. Every day my mother would send a list of the patients my father planned to visit to the operator at the exchange so that she knew where to reach him in an emergency.

The tranquility of Arnside is shattered twice a day by a tidal bore that roars up the estuary at precise intervals of twelve hours and twenty-five minutes. This leading edge of the incoming tide

FIGURE 1. Upper: Arnside estuary and viaduct. Middle left: Childhood home, Orchard Close. Middle right: The garden. Lower left: Living room. Lower right: Dining room.

is forced through the narrow entrance to the river Kent, transforming it into a wave of water only about three feet high but of unbelievable strength. The wall of water roars past Arnside, swirls round as it hits the viaduct, and splits into two, one current passing through the viaduct, the other curling round snakelike to charge in anger back toward the open sea. The strength of the current is such that horses caught in the bore have been knocked over and drowned, and there have been several human fatalities. Today, a siren warns people half an hour ahead of the incoming tide. When I was young, there was no warning; you had to know the times of the tides and how they changed each day. This meant being aware of the phases of the moon, which altered the strength of the gravitational pull, as well as the direction of the wind. We learnt that a small bore at half-moon produces a neap tide, and a large bore at new and full moon a spring tide. A ferociously powerful bore would occur when a strong west wind was behind a spring tide. A spring tide has nothing to do with seasons. The name is derived from the old Anglo-Saxon/German word *springan*, meaning "to leap forward." The name of the neap tide is derived from the word *nep*, meaning "lower." This awareness of our surroundings was essential to our survival as we had to resist the temptation to walk out onto the sands to look for stranded fish in pools, collect shells, or get a closer look at the waterbirds—turnstones, redshanks, curlews, sanderlings, shelducks, and Greylag Geese—that feed on the mudflats. Another danger was quicksand, which had accounted for the deaths of many dogs, a few horses, and even humans.

The danger of the bore, a habitual tsunami, was vividly impressed on us one calm summer afternoon in 1944. My brother John and I were perched on a dry tussock covered with Purple-flowering Thrift (*Armeria maritima*), waiting impatiently for the calm waters after the bore and before the turn of the tide, when we were allowed to swim. We had watched the wall of water displace the shelducks, curlews, and turnstones, which flew off screaming as the wave approached them. Simon, our springer spaniel, was paddling for flounders in the shallow pools, bringing them to my mother, who would later cook them for his dinner. Andrew, my three-month-old youngest brother, was asleep in his pram. Then, suddenly, we saw a child floating facedown in the stream of water swirling at speed back toward the open sea. My mother rushed into the fast-flowing current, managed to grab him and turn him upside down, holding him by his legs, and out of his mouth came a mixture of water, vomit, and seaweed. He regained consciousness while my mother was hooking out the remaining seaweed from his mouth, but he was shaking uncontrollably. We wrapped the shivering, sobbing boy in towels and put him at the bottom of Andrew's pram. My mother wheeled him as fast as she could up the steep hill to our house, urging John and me to help push, push. At the house, she told me to hold him, wrapped in a warm towel from the airing cupboard, while she filled a bath with warm water, telephoned the exchange, and told them to get my father, who was visiting a patient, and then the police to find the boy's parents. The boy, who was six or seven years old, recovered, and my father arrived shortly afterward, to my relief, and took over. Eventually, the boy's mother was found, shopping with another women, both totally unaware of the dangers when they left him to play alone on the beach. Even when they came to our house to collect him, they were still remarkably unconcerned. I was afraid they would scold him and hoped my father, who could be stern and very severe, would reprimand the mother before that happened.

Fast-forward forty-three years: Peter and I were hiking in the mountains of Nepal, high up in the Langtang Valley, with a guide, Pasang Sherpa (chapter 19). The three of us had been walking for several days. We stayed in huts (teahouses), the type where a wooden shelf for laying down your sleeping bag creates a communal bedroom, and smoke from an open fire used for cooking curls up through a hole in the roof. Pasang was talking to the owners, who were old friends of his. We had seen nobody all day, so we were surprised when two hikers from England walked in and joined us, sitting around the fire as we waited for dinner to be cooked. Chatting with them and exchanging information, I said I was born in the tiny village of Arnside, "which you have probably never heard of."

"Oh, yes, I have," the man replied. "When I was seven, I nearly drowned there and was rescued by the doctor's wife and her daughter!"

I missed the acute awareness of my surroundings when I was sent to boarding school in Edinburgh at age eight and no longer needed to perform the daily tidal calculations. They became important again many years later when we were camping on isolated and uninhabited Galápagos islands, and I had immense joy in once again being aware of the movements and timing of the tides, moon, and stars.

2

War

All war is a symptom of man's failure as a thinking animal.

—JOHN STEINBECK, *ONCE THERE WAS A WAR*

For most of my early childhood, Europe was engulfed by World War II. I was three when the war began, in September 1939, and eight when it ended, in May 1945. At the age of three, my understanding of the war was vague. Why were we being bombed? Why did people want to kill us? Are Germans men with germs? I asked. Who is Hitler? As I grew older, my parents would show me maps and explain what was happening. I knew that my father was helping Jewish families escape, but all details about how he did this were kept from me.

The war affected daily life in our village in two ways, by enhanced security and by food restrictions. German planes flew over us nightly on their way to bomb the Barrow-in-Furness shipyards. A potential target was Arnside viaduct, a long bridge that carried the railway line across the estuary, the only rail link between the south and the shipyards. No streetlights were allowed. My parents made blackout screens for all the windows. Each night they slotted them into the window frames, then one of them would walk around the house to make sure not a glimmer of light escaped. A room off the garage under the house was converted into a bomb shelter, with access down a rope-and-wood lift contraption from the kitchen. We were allowed out at night only when there was a full moon or enough moonlight to see our way. If my father was called out after dark to see a patient, he would either walk using a torch (flashlight) with a shade that projected a glimmer of light downward, or carefully drive the car with shades fitted over the headlights. Signposts were removed to confuse any invader. One on Arnside Knott had been overlooked, and so it was turned the wrong way. We were issued gas masks. John, being younger, had a Mickey Mouse gas mask, whereas I had a smaller version of the ugly adult one. We were taught to recognize bombs and flares that had washed up on the shore. These were quite common, and once, out on a walk with my mother, we found a pile of them at the back of the beach between Arnside and Far Arnside. Warning us not to touch them, my mother contacted the police, and the bombs were later detonated.

John became fascinated by the planes. He made models of them and hung them from his bedroom ceiling. To get from door to bed we had to circumnavigate and duck under de Havillands, Hurricanes, Hawker Demons, biplanes, seaplanes, gliders, Spitfires, and numerous others. Spitfires were our favorites; together we would watch them from the shore as they appeared above the trees and then dropped, skimming barely above the water until the very last moment, when they performed a neat hop over the viaduct. Both of us knew the sounds of them all, from the German bombers with their undulating deep-throated throb to the terrifying whine of the doodlebug, a bomb with wings like a plane, which prompted our parents to rush and throw us under the kitchen table, covering it with blankets to protect us from flying glass.

FIGURE 2. Upper left: My parents, 1935. Lower left: My parents, ca. 1953. Upper right: Brother John and Mother, ca. 1953. Lower right: Mother and brother Andrew, ca. 1953.

Arnside was spared the constant bombardment of attacks visited on more densely populated places. My father was concerned about his many friends who worked in hospitals and medical practices in the heavily bombed big cities, so he offered to exchange practices for three months at a time. A Dr. Hudson from Croydon, one of the worst hit areas in London, stayed with us on several of these exchanges, and we all became good friends.

In 1939, children were evacuated from urban to rural environments to protect them from the bombings. This was an emotional subject at home, because my mother was strongly against separating young children from their parents. She spoke from experience; her own mother had died when she was eleven, and my mother had been sent immediately to a boarding school, not even returning home for holidays. It was so traumatic for her that to her dying day she would not or could not speak of those years.

One day, sixty young children arrived in Arnside, where a building was being used as a center for distributing evacuees to foster homes. My mother rushed round to help and spent the day there. She returned distraught and burst into floods of tears as she walked up the drive to our house, the first time I ever saw her crying. She told me the room was full of small children, from babies to six-year-olds, some even separated from their brothers and sisters, all crying for their mothers. Even with her and several other women from the village trying to help, it was impossible to comfort them all. In a fury, she said she would never ever leave her children to such a fate.

Most of my friends as well as Peter, my husband, had been evacuees. They shared a common feature: they had successfully blanked out those years in their memory, or they refused to talk about them. Then, as happened with Peter at about sixty years old, many began to discuss their experiences. Books appeared on what it was like to be evacuated, and the stories they tell reflect a wide range of experiences, from positive to abusive, with sometimes lasting distress.[1]

Food was rationed, and this lasted until well after the war had ended: two ounces of butter, two ounces of cheese, four ounces of bacon, one egg, two ounces of sugar made from locally grown beet, two ounces of tea, and three pints of milk, per person per week. No sweets or candies. We were encouraged to keep hens, rabbits, pigs, and goats, if we had room, as part of the nation's efforts to be self-sufficient. It was forbidden to grow flowers; all flower beds had to be converted into vegetable patches. We were lucky to have quite a large garden and about half an acre of orchard. John and I helped my mother to plant vegetables and raise twelve hens, five turkeys, and two pigs. We stored apples in the attic on slatted wooden trays, so that they did not touch each other, and eggs in a large ceramic crock to which "water glass," or liquid sodium silicate, was added. The eggs became coated in white lime and lasted for about a year. This was useful because our hens did not lay many eggs during the winter. Since we had no refrigerator or freezer, our food was kept in a cool larder, a room on the north corner of the house with stone shelves that my mother whitewashed once a year with a type of paint made from slaked lime (calcium hydroxide) or chalk (calcium carbonate). During summer and autumn, we had an overabundance of vegetables and fruit. We would put them in boxes or in a wheelbarrow outside the gate for people to help themselves.

My mother, never completely conventional, had chosen hens that were a cross between junglefowl and Red Leghorn. They were basically wild and produced small but delicious eggs. Every night they tried to roost in the apple trees, and John and

I had the task of shooing them into their henhouse so they would be safe from rats and foxes. Years later, in the Terai in Nepal, Peter and I watched wild Red Junglefowl, which looked and behaved so exactly like our hens that I wondered if ours weren't close to pure junglefowl. Sometimes, one of our hens would disappear, and we would eventually find her under a bush, a broody hen incubating a clutch of eggs. Later she would rejoin the flock with a dozen or so baby chickens.

In 1944–46 my mother raised two Wessex Saddleback pigs. John and I called them Rodney and Priscilla, loved them, and spoiled them. We cleaned out their sty every day, brushed them with a stiff long-handled bristle brush, allowed them out into the orchard, and collected slops (left over potato peelings, etc.) for them in a bucket from our neighbors. When the time came to slaughter them, my mother made sure John and I were out of the house at school. She used every bit of the pig, even learning how to make sausages. Eventually, huge smoked hams in muslin bags and strings of smoked sausages joined the plaited strings of onions and bunches of dried herbs that hung from the kitchen beam. When we were served our first meal of pork, I asked, "Is this Rodney or Priscilla?" Then my little brother Andrew piped up, "Mummy, when do we eat Bracken?" Bracken was our family dog at the time. We sat down in tears. None of us, not even my father, could eat anything, so all of Rodney and Priscilla, the beautifully prepared sausages, large hams, and other delicacies, went to our very grateful neighbors.

My most profound and lasting impression of wartime was meeting some German prisoners. Several had been sent from a camp nearby to install a water main on the Arnside roads, one of which passed between our garden and orchard. Swinging on the garden gate, John and I would watch them, and one day we dared to cross the road to talk to them. The guard came up and

FIGURE 3. In the garden, 1944. Upper left: With my brothers, John and Andrew (I was wearing a dress my mother had made from an old sheet). Upper right: With my brothers, our gardener, Jeremiah Swindlehurst (Jerry), and our dog, Simon. Lower left: With my brothers and Elsie, our housekeeper. Lower right: With our pigs, Rodney and Priscilla.

said we could talk to them but only during their tea break. We waited for their break and rushed over. One of the prisoners showed me photographs of his children, a girl of my age and a boy of John's age. He hated the war and wished it would end because he longed to see them again. This was the first of many tea breaks we spent with the prisoners. This same man made a bracelet for me and a small toy for John. We thought this was our secret and that we had been talking to them without our parents' knowledge. After a while and feeling guilty, we broached the subject, only to find that they already knew. This dichotomy, being fearful of German bombs yet realizing these German prisoners also disliked war and were people like my parents with children like my brother and me, was very bewildering. My parents patiently explained about Hitler and his atrocities and how many men were forced to fight against their will. I knew my parents were helping Jewish people, although none of us knew at that time the extent of the horrors of the Holocaust. It was my first inkling of the power one evil person and his devotees could have over thousands and, by persuading his compatriots that the enemy is less than human, make war possible. I was six years old, and this vivid experience laid the foundation for my adult belief that communication between people from different cultural and ethnic backgrounds can bring a much-needed depth of understanding, respect, tolerance, and wisdom. At best, the rewards of an interchange of ideas are immense and can lead to imaginative and ingenious solutions to otherwise seemingly unsolvable problems, such as war.

3

Living in a Medical Household

Kindness is the language which the deaf can hear and the blind can see.

—UNKNOWN (OFTEN ATTRIBUTED TO MARK TWAIN)

Our daily life revolved around my father's medical practice. The practice constrained us, but I did not rebel against it; rather, I became partly incorporated into it. Initially, as young children, we were kept away from the wing of the house where my father received patients behind a soundproof door. But later, when I became older, I helped with the sterilizing of instruments, and even as a university student on holiday, I acted as an occasional receptionist and nurse if my father was doing a minor operation. It became impossible to be detached from the medical environment on our domestic premises.

A medical atmosphere permeated our social activities as well. Many of the inhabitants of Arnside were old, retired people, often living together. They told my father that they missed seeing children, so he would sometimes take John and me in the car on his rounds to visit them. We would wait in the car while he saw them professionally, and then we would be invited into the house. Many houses did not have electricity. We would be ushered into a shady room smelling not unpleasantly of a mixture of camphor, lavender, cedar, and paraffin. One very old lady had been a violinist in the Hallé Orchestra, and she played her beautifully polished violin for us. Mr. Dixon collected clocks. His house was full of them from top to bottom, different sizes and types all ticking out of synchrony with different rhythms, tones, and volumes. Being deaf, he was unaware of the cacophony and was slightly bemused that his two Scottie dogs preferred to sit by the goldfish pond in his garden rather than enter the house.

Frequent visits were made to the nursing home to talk with, and sometimes read to, the patients. *Yonderly* is a kind and sympathetic Westmorland expression for people who have lost their memory or are experiencing mental deterioration. We would ask, "How is Mrs. Smith?" and be told, "Very well, but she is a bit *yonderly* today." I often wondered if the patients' unique *yonderly* world, with its own logic, was superior to ours. Could they be making up poetry or stories like Lewis Carroll's "Jabberwocky" or "The Walrus and the Carpenter"? Were they much cleverer than we were? One lady was always sitting up in bed with her nightcap on, reading, sometimes with her book upside down. John and I would ask her how she was, and with a big smile, she would reply, "My dears, absolutely wopsie, absolutely wopsie!" We visited Mrs. Absolutely Wopsie on many occasions, until one day she disappeared into her own *yonderly* world and never returned.

Mr. and Mrs. Benson and Mrs. Benson's sister, Mrs. Leeming, lived together and asked us to visit them frequently. We did so for many years, even after Peter and I were married. When we were very young, they would give John and me tea, with scones and jam tarts baked in their iron oven next to an open fire. After tea they would take us for a long walk down to the marshes, where in springtime, if we were lucky, we would follow the sound of a bittern booming to find him camouflaged in the reeds. Then we would walk the three miles home, past a fossil-rich limestone quarry, through woods of coppiced trees, primroses, and wild daffodils, along the dikes with streams full of tadpoles and through the field downslope from the cemetery. This field was always green with particularly prosperous crops. "Was it fertilized by water from the dead bodies?" John and I whispered to each other. We dawdled sometimes until dusk, when we could hear the churring of nightjars and see pipistrelle bats. In winter, when it was too dark for the walk home, Mr. Benson would put us on the train at Silverdale, where the stationmaster lit the dim gaslights with a long taper precisely five minutes before the train was due to arrive. Arnside, only four minutes down the line, did not have gas, so my mother or father would meet us at the train with a flashlight, the only pinpoint of light in the otherwise dark station.

Twice I was caught up in household medical dramas. In the first, I was alone in the house when suddenly the double doors of the surgery burst open, and my father yelled, "Get the blanket off my bed and bring it here." I dashed into my parents' bedroom, dragged the blanket from their bed, bundled it up, and ran with it downstairs into the surgery. I stopped in the doorway and watched my father being chased by a stumbling, incoherent giant with a scalpel in his hand. My father was thin and not very tall, but nimble as he threw chairs in the man's way, going round and round the room. Then my father nipped behind him, kicked him in the back of his knee, and down he went crashing to the floor. My father grabbed the blanket from me, threw it over the man, and skillfully rolled him up, just as Tom Kitten is bundled away in Beatrix Potter's *Tale of Samuel Whiskers*. The man's head was sticking out at one end, and his shoes at the other. My father gave him an injection. I was told to leave, and shortly afterward the ambulance and police arrived. Later I found out he was after morphine or opium and that he had entered the surgery after the other patients, grabbed a scalpel from a tray by the small autoclave, demanded drugs, and threatened my father.

The second emergency was also dire but less dangerous. One morning toward lunchtime a van drove up our driveway, and a man clearly in distress called out. My mother and I rushed out and found him lying half in and half out of the van with blood gushing from a wound in his thigh. My father was out on his rounds visiting patients. My mother got me to help her cut his trousers and put a tourniquet round the top of his leg to stop the blood pouring from an artery, and then urged that it should be loosened every few minutes. She phoned for an ambulance, the nearest one being eleven miles away, and phoned the person at the telephone exchange, who knew my father's daily round and how to find him. I was told to stay with the patient. Fortunately, my father returned well before the ambulance arrived.

During the war, food was in short supply, and people hunted the numerous rabbits in the fields and woods or the occasional deer to supplement their rationed food. The man had been hunting rabbits all morning and was resting, sitting on the ground, not realizing that the barrel of his loaded gun was against an adjacent rock. His dog jumped onto his lap, touching the trigger, and the gun exploded, causing a piece of shrapnel to enter his upper thigh. Somehow, he managed to staunch the blood with a piece of cloth tied tightly round the top of his leg, crawl to his van, some distance away, and drive the two or three miles to our house. He was more worried about his dog than himself, because the dog had run away after the explosion, and he didn't know if it was lost or injured. All ended well, for the dog found its own way back home unharmed. The patient spent a few nights in hospital and eventually recovered with only a limp.

When I was five years old, another momentous event occurred that made a deep and lasting impression on me. When the waiting room was full, or a patient had a special need, my father would put patients in the dining room to wait, and there I sometimes encountered them on my way to the garden. That day, I was passing though the dining room, and sitting there to the left of the fireplace was a bent, shriveled-up old man. I murmured "excuse me," and he suddenly turned his eyes on me. I was transfixed, rooted to the spot, unable to move. He pointed to the bowl of lilies-of-the-valley on the dining room table that my mother and I had picked the day before. "Look, look," he said. "Do you not see, you must see, every flower is the soul of a man I shot and killed. They are all looking at me." He went on in this vein, and then his mood and theme abruptly changed. His eyes widened. "All I want is to be forgiven, and have a tiny house, with some hens, and live in peace. I am tormented." On and on he spoke. Suddenly, my mother appeared and took me into the garden. Afterward, my parents explained that he had fought in the trenches during the First World War and was suffering from shell shock, and that my father was trying to help him.

Years later, when I was at school, we were given Samuel Taylor Coleridge's "Rime of the Ancient Mariner" to read. The memory of my encounter with the poor soldier came rushing back to me, and I felt I could understand the poem more deeply than my teacher or any of my classmates could, but I did not have the capacity to put into words what I felt for fear of breaking down in tears.

The poem begins with my experience:

He holds him with his glittering eye—
The Wedding-Guest stood still,
And listens like a three years' child:
The Mariner hath his will.

Later in the poem:

And I had done a hellish thing,
And it would work 'em woe:
For all averred, I had killed the bird
That made the breeze to blow. . . .

Four times fifty living men,
(And I heard nor sigh nor groan)
With heavy thump, a lifeless lump,
They dropped down one by one.

The Souls did from their bodies fly,—
They fled to bliss or woe!
And every soul, it passed me by,
Like the whizz of my crossbow! . . .

Alone, alone, all, all alone
Alone on a wide wide sea!
And never a saint took pity on
My soul in agony.

The many men, so beautiful!
And they all dead did lie:
And a thousand thousand slimy things
Lived on; and so did I. . . .

The final two verses:

The Mariner, whose eye is bright,
Whose beard with age is hoar,
Is gone: and now the Wedding-Guest
Turned from the bridegroom's door.

He went like one that hath been stunned,
And is of sense forlorn:
A sadder and a wiser man,
He rose the morrow morn.[1]

4

Cooperation and Respect

The only thing that will redeem mankind is cooperation.

—ATTRIBUTED TO BERTRAND RUSSELL

A concept prevalent in the north of England discourages anyone to think they are superior to anyone else. This egalitarian belief was thought to have been brought to the region by Norse shepherds when they first settled in the Lake District. They emphasized cooperation rather than competition, for how else could they so successfully develop and maintain the idea of common land for tending their Herdwick sheep for over a thousand years? This precept was held and applied rigorously by my mother.

My mother laughed a lot; she was fun but firm and rarely angry. Usually our misdeeds were strongly corrected, but we knew she really thought of them as mischievous pranks. Twice she was genuinely angry with me, and both incidents involved a lack of empathy on my part. The first occurred in the year after the war ended, at a time of increasing food shortage. Two young

German men, I imagine in their late teens or early twenties, arrived in Arnside on bicycles. My mother and I were shopping in the village, and all she managed to buy was one loaf of oat bread and a half dozen eggs. It was the time of year when hens were not laying, and there was very little food in the shops. These two young men asked her if there was any place they could buy food, any food. She told them of two possibilities, warning them that there was not much food available. Then she gave them our address, saying, if you cannot find any food, this is where we live—come and ask me.

Before we even entered the house, they arrived in our driveway. They had not found anything. My mother gave them half of the loaf of bread and two eggs. After they had left, I said, "But Mummy, they are German." Then the storm broke. "You have had breakfast and lunch, and they have had nothing all day. People are people. This has nothing to do with Hitler. The war is over; they are two boys who are very, very hungry."

The other outburst came when I was eleven. I had humiliated the person who came each day to help my mother with the housework. Her name was Elsie. She was small, a very pleasant person with a humped back and limp caused either by polio or a developmental problem at birth. She had never been to school and could not read or write. I corrected her spoken grammar, which clearly made her uncomfortable. My mother took me aside, out of Elsie's hearing; she was furious. She delivered a long, stern lecture about how fortunate I was, whereas Elsie and others were not, and so on. "Don't ever think for a moment that you are superior to anyone," was the clear and unforgettable message.

Elsie was a kind, loving person. She adored my younger brother, Andrew, who was a baby at that time. When he was older, to her delight, he became her doctor. Peter and I last saw her on one of our visits to England when she was living in a little suite in a retirement home in Arnside. She was over ninety and suffering from cancer but extraordinarily happy. She loved her little room with a view over the estuary and had recently found the joy of audiobooks. A free library delivered six books of her choosing each week. She had asked them not to bring "romances." She preferred books from the section marked classical—"Whatever that means," she said—adding, "especially when read by a man with a deep voice." That day she was listening to a book by Charles Dickens. Never having been able to read, she was relishing a whole new world that had suddenly opened up for her.

Both my parents taught us to value the people around us. Jerry—Jeremiah Swindlehurst—was my favorite, and I think of him as one of the people who inspired me to become a biologist. He helped in our garden during and after the war, when all garden beds had to be devoted to growing vegetables. Like Elsie, he had never been to school, but he had taught himself to read and write and liked nothing better than to listen to the BBC Home Service (now Radio 4), a program of classical music, serious drama, literature, and discussion. I remember him talking to my parents about how much he enjoyed the Reith Lectures, a series inaugurated in 1948 to enrich the cultural life of the nation. Before Jerry died in the late 1950s, he must have heard lectures from people such as Bertrand Russell, Arnold Toynbee, and Robert Oppenheimer.

Jerry was a wizard in the garden. His wisdom, curiosity, and knowledge about plants and animals appeared to be endless. He showed me the eggs in a bird's nest, telling me why birds nested where they did, why their eggs were colored, where dormice and glowworms could be found, and a host of other facts about nature. He taught our family all the tricks of gardening, which vegetables to grow next to each other, and how to rotate crops between years to get the best nutrients for them and avoid pests.

After the war, Jerry had saved enough money to buy three greenhouses and spent his days growing vegetables that he sold to the local shops. We would often visit him there and sometimes call in at his house if we were passing. On these occasions, his wife would open one of her many cake tins and offer us a small slice of seedcake. She had a seemingly endless supply of seedcake that she kept in tins piled high on a shelf, almost to the ceiling. I think she and Jerry collected caraway seeds, then made the cakes, flavoring them with mace, nutmeg, and quite a bit of brandy. Adding brandy at intervals apparently preserved the cakes for years, to judge by the number of cake tins she had. I did not like the cake. Knowing it was considered very special, I surreptitiously handed my bit to my mother when nobody was looking.

5

The Dining Room Table

I cannot teach anybody anything. I can only make them think.

—SOCRATES

The dining room table was a focal point for mealtime conversations between adults and us children. It was circular, six feet in diameter, and made in Burma (Myanmar) of three types of tropical wood—rosewood, ebony, and teak—arranged in spokes. Laughter, quarrels, and lighthearted and serious discussions circulated around it during our daily meals. I learned much from the serious conversations, shown by parental example how to be critical, skeptical, and ethical.

Our house was like a magnet, with people constantly coming and going, frequently and without warning staying on for lunch or dinner. My mother would quickly make up a meal with eggs, a tin of beans, or any excess vegetables on hand. Our visitors then joined us around the table, adding to the variety of opinions. One special guest stayed with us for a week. This was Mr. McIntyre, a tall, thin, stern-looking man who had been my father's classics schoolmaster. He had stimulated my father's love of classical Greek and Latin and had also encouraged him to go to Glasgow University. My father was the second-youngest of twelve children, and the only one to go to university. Many people, my mother included, made their own clothes during the war, but Mr. McIntyre outshone us all. He collected sheep wool from barbed-wire fences, carded it, dyed it with his own mix of vegetable dyes, and then wove the cloth and made his warm three-piece wool suit of muted watercolors.

This was the first time I saw my rather domineering father become submissive; he was in awe of Mr. McIntyre, and as a result so were we. He was placed between my father and me at the table. At our first meal with him, we had one lamb chop each, and this was very special, as food was hard to come by. I turned to him and said, "Are you family?" He gave me a startled look, and I had a fleeting vision of being a pupil in his class. I quickly added, "If you are family, I am allowed to pick up my chop in my fingers and gnaw the meat from the bones, but Mummy said not to do this in polite company." Roaring with laughter, he promptly picked up his chop with his fingers. From then on, all formalities ceased, and he and I became the best of friends.

During the week, he told me about the Lyceum in ancient Athens, a university that women as well as men could attend. He said it was surrounded by statues of nine beautiful women, the Muses that inspired creativity in the arts and sciences. As a child, I particularly liked the idea of statues of women, as most statues in England were of tough military men or kings on horseback honored for their heroism in wars. Mr. McIntyre talked about how Eratosthenes calculated the circumference of the world by measuring the angle of the shadow cast by the sun at noon on June 21 each year, in different locations. It seemed so simple, and thinking of the Castlerigg stone circle near Keswick, constructed around 3000 BCE, and the ones at Stonehenge,

I wondered whether other people long ago had done similar calculations that were never recorded.

On another occasion, we had just returned from church and were having lunch, when my father turned to John and me, age four and six, and asked, "Well, how did you like the sermon?" "Boring," was our instant reply. "Well, how about the miracles? What about Jesus walking on water?" "Impossible," we said. The children's encyclopedia was kept in the dining room, and we were told to look up the salt concentration of the Dead Sea. Then, pouring salt into our glasses of water, we started floating objects on the surface and comparing them with those that sank. We decided the Dead Sea might be salty enough for a body to float but not walk on. Then my father told us about mirages and showed us, with a prism and a mirror, how light waves can be bent by refraction, and explained how this optical illusion can occur if the air below was cooler than the air above, which was a possibility over the Dead Sea.

Next, came the question of the miracle of the burning bush; could spontaneous fires ever occur in oil-rich regions? According to the encyclopedia, yes, they could, especially when ignited by lightening. If I had known then what I do now, I could have contributed the amazing example of the synchronously flashing fireflies that, by densely covering a single bush, can give the impression of fire. Miracle number three, feeding five thousand people with five loaves and two fish, was a problem until my mother said, "You know how in the summer we put out vegetables and eggs for anyone to take? Well, that encourages others to do the same. Quickly, you can accumulate enough willing contributors to feed five thousand!"

Shortly after that conversation, my mother was visited by the vicar, who called her a Doubting Thomas. The attitude of my parents toward religion was that the Bible should be viewed as allegorical and symbolic. My mother was extremely upset when the poetic King James version of the Bible, which she thought encouraged metaphorical interpretations, was replaced by the modern literal rendering.

As we grew older, the topic of religion turned into a discussion of how all the world's religions have the common themes of tolerance, compassion, forgiveness, and charity. Yet, all religions across the ages have been used to rally people to war in the name of their favored God. Even Buddhists, thought to be a model of tolerance, had their warrior monks in ancient Japan.

In the last few days of my father's life, when he was eighty-eight, frail, and aware that he was dying, he took up the theme of science and religion again and wrote a short essay entitled "COSMOS." I quote from his final paragraph:

> *I do not wish to decry either Science or Religion. In the light of our ontological unknowledge it does not seem so terribly important what one believes, or what one of the amazing varieties of religious beliefs one holds, as long as there is a belief and practice in a code of ethics. . . . We have only the one earthly world to inhabit, and that for a period of a mini-second of biological time. All our various beliefs and disputations that are at war with each other or in disagreement over doctrinal minutiae will fade into insignificance if we neglect these wider issues. We should be striving to preserve our world in at least as presentable a condition as that in which we found it and freedom of opportunity which we have always sought for ourselves. . . . To strive, to seek to find—which is the brief of Science, is not necessarily amoral, and its findings, if morally applied, have the potential for a better world—and that here and now . . .*

The final section of his essay ends thus:

> *A Code of moral behaviour of universal acceptance and practice is to be desired. Much of this is already in many of the chief and established religions of the world but obscured by the emphasis on the metaphysical beliefs which separate them. Charity (loving-kindness if you will) and concern for all people and our environment are the greatest virtues that stand above all others and are their own reward.*

Three days later, he died. I reflected on my father's final words in 2009, when Peter and I were greatly honored and humbled to be awarded the Kyoto Prize. In the words of Kazuo Inamori, who founded this generous prize, "A human being has no higher calling than to strive for the greater good of humanity and the world." His sincere wish was "to contribute to the progress of the future of humanity while maintaining a balance between the development of science and civilization and the enrichment of the human spirit." How I wish my parents had still been alive when Peter and I received this wonderful award for our research and contributions to knowledge, as they would have had the opportunity to meet Kazuo Inamori himself.

6

Fossils in the Fire

Wonder is the beginning of wisdom.

—SOCRATES, AS REPORTED BY PLATO

For most of my life, I have been fascinated by questions about the past and how the world has changed. Where did this interest begin? It probably began with the discovery that fossils are the remains of animals and plants that lived eons ago. Fossils stimulate us to imagine another world, distant yet connected to ours through time. I can remember the exact moment and place when I was first captivated by this realization. I was three or four years old and staring into a fire.

In our winter routine, John and I would play in the bath while my mother warmed two towels on the guard in front of the fire. Then she carried us, one by one, wrapped in a towel, to have our supper in front of the fire before taking us to bed, where she would read us a book. John, being younger, was first. One night, sitting in front of the fire while waiting for my turn, I saw a magnificent fossil fern in a shiny lump of coal, soon to be burned and extinguished forevermore. I rescued lump after lump of these fossil ferns from their incandescent fate. I had seen fossil shells—for example, those of gastropods and brachiopods—and fragments of corals on the beach, in our garden, and even in the limestone on Arnside Knott, but they paled in comparison to these exotic burnished ferns, with every frond and spore etched in the polished coal. In addition to ferns, there were club mosses and horsetails (*Equisetum*). I was familiar with many species of ferns and horsetails, with their jointed stems and whorls of minute modified leaves, which grew as weeds in our garden, but as my mother explained, these living forms were tiny descendants of the magnificent trees that had grown to thirty or forty feet high very long ago. Soon I had a pile of fossil coal specimens on the edge of the fireplace that were not allowed to be burned. When we were older, John grumbled that if I went on like this, we would soon get very cold.

Standing on the top of Arnside Knott, overlooking the estuary, my mother would ask, "Can you imagine what this was like before people lived here?" Both my parents played this game with us. It was always fun, just a game; nevertheless, these early questions stimulated a lifelong interest in the past and how the world has come to be the way it is. It was the first time I realized people were not always on this earth. I thought back to my fossils. How long ago were those fossils living organisms, and why were they so beautifully preserved? A huge fossil coral was in our garden. How did it get there? Had Arnside really lain on the edge of a warm tropical sea with corals, its shorelines flanked by giant ferns and horsetails?

Later, when my geography teacher introduced me to the possibility of whole continents drifting apart (chapter 7), I was excited to discover we would have to jump back 300 million years to the Carboniferous period to reach this idyllic setting. With her help, I was able to piece together a possible time line for Arnside. The history of Arnside spans an enormous time from the formation of the mountains by volcanic activity in the Lake District 490 to 390 million years ago to the last glacial-interglacial cycle approximately 20,000 years ago, which carved out U-shaped valleys, created the Lake District's characteristic narrow ribbon lakes, and pushed sand and gravel into low, oval-shaped hills called drumlins, common in the fields around the village.

When did humans first appear, and what impact did they have on the world? These were the questions I asked my parents in response to their game of imagining a prehuman time. There were rumors that the earliest evidence of humans in the Lake District had been discovered in Kirkhead Cave, on the opposite bank of the river Kent from Arnside. In this cave, human bones were found together with bones of Moose (*Alces alces*; known as Elk in Britain), Wild Horse (*Equus ferus*), and Gray Wolf (*Canis lupus*). These remains were later dated to be more than ten thousand years old.[1] Pollen cores from Hawes Water, a lake near Arnside, gave insight into the vegetation at this time: dwarf birch (*Betula*), willow (*Salix*), and juniper (*Juniperus*).[2] I liked to imagine the human hunters stalking their prey through this shrubby landscape. A layer of charcoal in the pollen cores suggests they cooked over fire. These ideas of ancient Brits, as we called them, were incorporated into the games John and I played.

My mother took us to the Castlerigg stone circle near Keswick, thought to have been built fifty-two hundred years ago. The purpose of this monument, Stonehenge, and others throughout Europe has been much debated. Some think they were meeting places for trade, whereas others point to the stones' alignment with lunar positions or midwinter sunrise. Alexander Thom, a mathematician, argued that the builders understood the lunar and solar cycles and arranged the stones as a calendar.[3] I tend to favor his interpretation, since my life in Arnside was governed by the need to understand the phases of the moon, to know when the tidal bore was due and when it was safe to swim; surely a knowledge of the seasons and phases of the moon would have been important to a farming and fishing community whose members needed to know when to plant and harvest food. Later I discovered that the use of stone circles or equivalent as calendars was widespread in early societies in South America as well as in Europe (chapters 13 and 17).

What did these people look like? Where did they come from? What language did they speak? I was intrigued to find that some place-names in the Lake District are remnants of an early language. *Cumbria* comes from *Cymry*, meaning "fellow countryman"; *Morecambe* Bay from *mori*, "sea," and *kambo*, "crooked"; river *Kent* from *cunetio*, "sacred one"; *Derwent* from *derwentio*, "oak valley"; and *crag* from *carreg*. These names are like fossils, remnants of an extinct, early Celtic language called Brythonic. I became fascinated as a child with the Brythonic numbers, which shepherds still used to count their sheep, and I learned to count in the Westmorland way. The numbers from one to twenty are: *yan, than, teddera, meddera, pimp, settera, lettera, hovera, dovera, dick, yan dick, than dick, teddera dick, meddera dick, bumfit, yan-a-bumfit, than-a-bumfit, teddera-bumfit, meddera-bumfit, jiggot*. A shepherd counts to twenty, puts a pebble in his pocket, and repeats the process; five pebbles equal one hundred.

Sheep rearing is omnipresent in the Lake District. In the ninth century, farmers originally from Norway arrived with their Herdwick sheep. There is no evidence for anything other than a peaceful coexistence with the Celtic farming community, possibly through intermarriage. Both groups were likely to have been small and widely dispersed and to have maintained mountain sheep for their livelihoods. A legacy of Norse names for villages, mountains, lakes, and streams remains to this day; for example, *beck,* a stream; *foss,* a waterfall; *ghyll,* a ravine; *howe,* a hill; *tarn,* a small lake; *thwaite,* a clearing; and *kirk,* a church—younger linguistic fossils.

The Herdwick sheep are tough and wild, and I wondered how much they had changed in behavior and appearance from their wild ancestor, Mouflon, a species of southwest Asia. A Herdwick sheep imprints on the patch of fell it was associated with early in life, and as a result, it will not wander far and so requires little herding in summer. Just as in the tenth century, each farm has a "fell right," which permits farmers to graze a certain number of their sheep on common land. Only when they are brought down to winter in fields in the warmer valleys do the sheep need herding, and then well-trained sheepdogs are a necessity. This system has prevented overgrazing for over a thousand years. It is threatened today by a Ministry of Defence policy to de-register some of the large, upland common fell lands and turn them into private land. This is the greatest threat to fell farmers since the similar enclosure acts in the eighteenth and nineteenth centuries, which allowed wealthy people to buy up the common land, forcing farmers to become serfs on their own land or leave to find work in cities. Fortunately, 20 percent of the Lake District is now a UNESCO World Heritage Site, thanks to a donation from Beatrix Potter to the National Trust. The remaining land is in jeopardy.[4]

This is the environment on which I, like the sheep, imprinted, and I share their connection with history. Questions concerning the movement of people and their genetic and cultural influences on different communities never left me, and throughout life I find I am drawn to articles on the subject. They even haunt me at night. Since childhood, I have had a recurring dream of walking back through history. The last one was only a few weeks ago. I started out in the town of Kendal. The shops, houses, and most people gradually faded away, leaving only a few tiny, scattered wooden and turf farmhouses. Suddenly I realized Peter was not with me, and in a panic, I explained my plight to a woman emerging from one of the houses. "Oh, that's easy," she said. "Just walk back the way you came for five thousand years, turn left, and he is in the car in the parking lot." I rolled to the left, and there he was, asleep in the bed beside me. Panic over!

7

School

> Imagination is more important than knowledge. For knowledge is limited to all we now know and understand, while imagination embraces the entire world, and all there ever will be to know and understand.
>
> —ALBERT EINSTEIN

Schools can be centers for creative and innovative learning or, as was sometimes stated in the 1940s, for molding a character to conform to society. My first school was neither, my second school both but not in equal proportions. At four years old, I wended my way through ginnels (narrow passages between high stone walls) to a small one-room school where Miss Lishman, the only teacher, had control of ten or twelve pupils, boys and girls, spanning the ages from four (me) to twelve. We drew and painted but were largely left to ourselves. I learned much more from my parents. I left the school in 1945, after the war ended, when I was eight years old. My parents decided to send me to a girls' boarding school in Edinburgh, Saint George's School, as there was no strong academic school for girls nearby. This was the era of strict segregation of the sexes, and schools beyond the age of seven or eight had either boys or girls, not both.

I dreaded my loss of freedom and confided in Jerry. Expecting him to be sympathetic, I was stunned when he turned toward me, with a touch of anger that I had never seen in him before, and said, "You are very lucky." I suddenly realized how deep was the loss of opportunity that he and others like him had experienced because of a lack of education. A few hours later he came and told me he would plant a tree for me that would last for many years, and together we planted a tiny spruce tree. Unfortunately, it died during my first year at school, so we planted another. From a picture on the internet, I see this one is now huge and still flourishing, almost eighty years later. The metaphor of death and replacement turned out to be more apt than either of us imagined.

Cities can be lonely places, despite, or because of, the immense number of people. I realized this on my first visit to Edinburgh. Crowds of people, all bundled in coats against the cold east wind, cocooned in body and mind, jostled each other as they walked fast down the streets, looking neither left nor right, each apparently determined to reach their own destination on time. This was so different from the cheery greetings of passersby, strangers or not, I was used to while walking in the village, through woods, or over the fells. In place of the fresh winds blowing over the estuary, here the atmosphere was smoky and gray, reeking of burning coke. My mother's sister, Auntie Jo, lived in Edinburgh and looked after me when we were allowed out every second Saturday. She proudly showed me the Forth Bridge, with its barrage balloons still flying to protect it from the German planes; the hill called Arthur's Seat, brooding over the city; and the forbidding-looking castle, with its dungeons and cannons, one of which was fired at one o'clock every day so that ships in the Firth of Forth could set their clocks.

I spent the first night at Saint George's in a small dormitory with two other new girls, who were so terrified they could not speak and cried for their mothers. Soon we met other girls and made friends, but between pupils and boarding matrons there was an us-versus-them attitude. This was another shock, as I had been used to mixing and talking with people of all ages. Nevertheless, I soon joined in the fun of interacting with other girls, sneaking flashlights into the dormitory so we could read in bed or, if one of them was confiscated, arranging the dressing tables so the light from the few remaining torches could bounce off the mirrors, allowing all ten of us in the dormitory to read. I had been used to a bath every night and a change of underwear each day. Here, baths and clean clothes were strictly regulated, occurring twice a week, and baths for us small children were shared, three in a tub, extending the still allotted wartime ration of four inches of water per person.

Punishments were severe. After infractions, such as talking, playing, or reading after lights out, Saturday outings were canceled, and instead we had to sit in and sew. Once we were given pillowcases to sew for a prison. I wished the prisoners well. Sometimes, the punishment was a spoonful of castor oil, which we quickly learned to hold in our mouths until we could get to a lavatory to spit it out, otherwise the stomach pains were excruciating. We had long discussions together about whether it was better to be beaten, as boys were in their schools, or suffer these abuses. My first punishment came soon after I arrived, when we were given black pudding for breakfast. Knowing it was made of congealed pig's blood, I refused to eat it. So, back came the same plate at lunch, then dinner, then breakfast the next morning. At that point, hungry and furious, I escaped, ran to a phone box, and illegally dialed my parents. "Take me away from here!" I am not sure what happened next, except there was no more black pudding for me, and I stayed at school. One of my friends did escape, saving up her pocket money for the train fare, running out of school, and reaching home, only to reappear the next week.

At this young age, we clearly required parental guidance, much as we considered ourselves to be independent. Only a small fraction of the pupils at the school were boarders, as I was; the rest were day girls living at home. The dichotomy between us grubby, smelly boarders and the clean, bright, soap-scented day girls became increasingly apparent as the term progressed. The day girls not only smelled better but did better at their schoolwork. We were left to our own devices to learn spelling, write essays, and hand them in on time or not. Our parents would have insisted that we did our homework had we been at home. Instead, we banded together, climbed an old elm tree on the grounds, made up plays, and generally had fun most of the time, while avoiding excess schoolwork and neglecting all homework.

Childhood diseases spread like wildfire among the boarders. I seemed to have a new cold once every three weeks throughout winter, but the worst was measles. Another girl and I came down with it together and were confined to the sickroom with a young and kind matron. I remember days blanked out with a high fever and intense headache. After we recovered, we had to convalesce for three weeks. A volunteer woman from the church came and took us for gentle walks, increasing the distance each day. All this added up to six weeks of missed schooling. At age ten that did not make much difference, except for mathematics, which I had previously enjoyed because I found the logic easy to follow. Now I was a long way behind. No one thought of coaching us through this period, and I did not ask for help. It would be two years before I finally had a breakthrough and caught up.

The attempt to shape our personalities began early. The school was strictly religious, following the Scottish Presbyterian, Calvinistic doctrine founded by John Knox, whose sixteenth-century house we were shown on the Royal Mile. We went to church each Sunday in crocodile formation with hats on and clean underwear and listened to the pulpit-thumping minister telling us we were all born sinners, having inherited Adam's guilty apple. Each night our house mistress would reiterate these messages and read the Bible while we were gathered in her sitting room darning holes in our socks. Even if the message was cold, at least her sitting room was warmer than the rest of the house, which was always freezing. Winter nights were so cold that the soap would freeze to the wash basin, and the water in the toilet would be ice the next morning, so we took it in turns to be the first to pee and make a hole in the ice.

We were expected to conform to an image of superiority that was in sharp contrast to my parent's egalitarian attitude toward all people (chapter 4). If we behaved like a queen, we were told, we would be treated like a queen, as if we all wanted to be one! On our days out, we were not allowed to enter Marks and Spencer or Woolworths, only the upper-class shops, like the expensive Jenners and Forsyth's. That made Woolworths and Marks and Spencer sound exotic, and it is where we headed on our Saturday outings after making sure there were no teachers in sight. Letters home were written on Sundays after church, and when we were young, they were vetted for spelling and content by the house mistress before they were sent home. This was the beginning of my weekly handwritten letter home, which continued until my mother died in 1995.

When away from them, I enjoyed writing these weekly synopses of stories, thoughts, and amusing events, and I looked forward to their return letters. It was a communication skill that did not translate well to the computer, where rapid exchange enforces brief, concise messages.

Not all was doom and gloom. There were instances of great kindness, and two of the teachers encouraged our creative curiosity. The first instance of kindness occurred during my first lunch in the large dining hall of the main school building. We were divided into houses; I was in Buccleuch. The head of the house was a tall, elegant girl with skin the color of glowing mahogany. I was the youngest at the table and was told that as the youngest, I had the task of getting the jug of water from the kitchen. The dining room, with its long tables arranged in rows, seemed to me to be endless, and the kitchen was somewhere, but in which direction I had no idea. It was pointed out to me, and off I went, passing the high table with all the mistresses staring down at me, and eventually found the kitchen. To my horror, the jug was huge and full to the brim with water. I bent down and heaved it up, wondering how I could possibly carry it without spilling, when suddenly the head of house appeared and carried it for me. She was delightful and kind, bubbling with good spirits. I always tried to sit as near to her as possible at lunch. This was her last year, and she left in July. Three years later I heard she had gone to Florida and while there had committed suicide. I was devastated. This was my introduction to the appalling racism and slavery in all but name that is still with us today.

The excitement of new ideas and experiences eclipses all attempts to force a child to conform. Two teachers stood out in this regard. One was the ballet teacher, Miss Middleton, who had taught at the Bolshoi. We did our standard barre exercises to music, but after that we were encouraged to choreograph our own dances while Miss Middleton imperceptibly introduced us to the intricacies of rhythm, composition, music, gesture, and dance. I became good friends with Miss Middleton's daughter, Christine. She played the piano beautifully, as well as several string instruments including the violin and cello, and composed music. Later she trained in Italy and became an accomplished public performer.

Few teachers have the courage to ask questions that they themselves cannot answer. Miss Crawford, the geography teacher, was one. Her questions and stimulating discussions sent us to maps and articles in her huge library. They were arranged on shelves beneath a collection of ostrich eggs, fossils, and a porcupine quill. I was eight years old when I first met her. I was puzzled by how you could be colder on top of a mountain than at the bottom, when you are nearer the sun at the top. Here she did know the answer. She tried to explain that it was because of atmospheric pressure, something about air particles being squashed together, low down, colliding and producing heat. All this was too confusing and above my eight-year-old head. She said, "I will show you." Two other boarders from my class also volunteered; we were all longing to escape into the outside world. She told us to creep around the school, duck as we passed the headmistress's window, and get into her little brown Morris Minor car. She had most likely told the headmistress she was going to take us, and this was to make it a thrilling and daring adventure we would never forget.

She drove to the nearby Braid Hills, grassy, low remnants of Devonian volcanoes dotted with clumps of gorse bushes. Although the hills are only about seven hundred feet high, that was enough elevation to reveal the drop in temperature recorded by the thermometer Miss Crawford had brought along.

Returning to school, we stopped at a small stream that was, she explained, a tributary of the massive river Forth. Then, hanging over a bridge of a major highway, we watched lorries (trucks) carrying goods from the port of Leith docks. She told us that the oranges were from Israel, bananas from the Caribbean, and spices from the Far East, packed in sacks made of jute from India. In this way we were connected to the outside world. I remember her laughing as she drove the car and saying, "Rosemary, I can see you in the mirror; you can't stop smiling."

On rainy days, we would crowd into her room after lunch, when she would put on films that transported us to other countries and the ways of life of animals and people. The flickering gray-and-white images accentuated the mystery and magic of peering into the unknown. One film was *Nanook of the North,* which gave insight into Inuit peoples' way of life. Another was of salmon streaming up rivers in British Columbia to lay their eggs, with bears lying in wait on the banks and then wading into a turmoil of spawning fish to scoop them up. Yet others depicted life inside a beehive and on an otter slide.

History or geography? At one point in my school career, we had to make a choice. We could not do both. History for me was rote learning a series of dates and names of kings, queens, and battles. I hated wars. In contrast, geography was full of thought-provoking questions. I was amazed how few people chose geography, which was lucky because it meant I was in with a small group of eight curious-minded students who loved discussions. Miss Crawford's lessons were like no other. An example was one about continental drift. She explained it was a theory proposed by Alfred Wegener but not accepted by the scientific establishment because he was unable to explain the mechanism of the movement of continents. She asked us to draw and cut out the continents and fit the pieces together like a jigsaw puzzle, then to draw in the geological rock formations and fossils that crossed continents. The results were compelling, but there was still the niggling problem of what made continents move. This led to long discussions. In 1968, long after I had left school, Bryan Isacks, Jack Oliver, and Lynn R. Sykes gave support to Wegener's theory by providing a mechanism: plate tectonics.[1]

How successful was Miss Crawford's unconventional approach to teaching in a conventional society? It was fun and thought-provoking, but did it work? When the time came for us to take the leaving certificate exams, which are given to all Scottish schoolchildren at sixteen years of age, the girls in the history group were cramming for their exam. I panicked. "Miss Crawford, we have only two weeks to go. How do we revise for these exams? We have no books, only papers." She said, "Just make sure you know the names of major rivers, mountains, and cities, and you will be fine. All you need is an atlas; the rest is in your head." She was correct; we were the top eight in the whole of Scotland that year.

All good schoolteachers take a personal interest in their pupils. Bullying was rampant at Saint George's. I was frequently called a "Sassenach," a disparaging term for an outsider or English person. I could shrug off the annoying implications, but much worse was the unrelenting tormenting of two girls who struggled with learning. All the teachers ignored this except for Miss Crawford, who took the whole class into her room to explain, as my parents had done, how to value all individuals. She pointed out the qualities of the two sufferers, saying how we could all learn from them. The difference this fifteen-minute talk made to classroom harmony was astounding.

As I grew older, Miss Crawford lent me books. One, *Seven Years in Tibet*, by Heinrich Harrer,[2] gave me a strong desire to go to Tibet, which Peter and I eventually accomplished many years later. Knowing my love of music, she arranged for me and a couple of friends to escape from the boarders' house to hear the Scottish National Orchestra. The concerts were in the Usher Hall, where we sat in "the Gods," at the very top, almost touching the ceiling, the cheapest seats available but with phenomenal acoustics. Other treats were Royal Scottish Geographical Society lectures and the occasional biological talk given by a professor from the University of Edinburgh. The two most memorable were one delivered by Edmund Hillary, shortly after he and Tenzing Norgay had reached the top of Mount Everest in 1953, and another about the regeneration of limbs in lizards that explored why reptiles can regenerate tails and limbs while mammals cannot.

Apart from geography, the academic level was not challenging. I daydreamed my way through many of my lessons. English literature was Jane Austen, Jane Austen, Jane Austen. To me her books were Regency soap operas, her heroines moving up the social ladder and out of genteel poverty by marrying a wealthy member of the landed gentry. Her books made little mention of how these landowners acquired their money. Instead of Austin, I devoured Charles Dickens and then Leo Tolstoy, Fyodor Dostoevsky, Joseph Conrad, George Orwell, and many others, by flashlight under my blankets at night. Two banned copies of *Lady Chatterley's Lover*, by D. H. Lawrence, probably the 1932 expurgated version, surreptitiously emerged from hiding places behind the bookcase in the boarders' house and circulated among us all.

During my allotted day out every second Saturday, I met three people who had a lasting influence on my career. Auntie Jo worked at the Poultry Research Center next to the Institute of Animal Genetics at the University of Edinburgh. She introduced me to her friends Peter Medawar, a future Nobel Prize recipient who discovered the acquired immune system through his work on skin-graft rejections; Charlotte Auerbach; and Douglas Falconer. Auerbach had written a children's book, *Adventures with Rosalind,* under her pen name, Charlotte Austen, and gave me a copy in 1947, just after it was first published.[3] I was a bit old for the book at the time, but I read it to my own children and realized what a delightfully imaginative book it is. A Jewish biologist, she had escaped Nazi Germany. At the time I first met her, she was working on the mutagenic effects of mustard gas, a particularly nasty compound used in chemical warfare in the First World War. Geneticist Douglas Falconer showed me his mouse colony and explained how he had been selectively breeding for large and small sizes and various mutants with neurobiological problems. He described how a large amount of genetic variation was required for evolution to occur by natural or artificial selection. Years later, when I was a student at the university, both Auerbach and Falconer were among my professors. Their lectures made a lasting impression; the knowledge I obtained from them was fundamental when conducting my own research.

Holidays at home did not come around fast enough. As I stepped off the train at Arnside station, the salty smell of the sea, the bubbling calls of the curlews, and the unique watercolor light would bring me almost to tears. My mother and I resumed our long walks, identifying wildflowers and watching birds. In the evenings I would curl up with her to listen to music and her explanations of the intricacies of the pieces. When I was twelve, on one of these holidays, my father said, "If you like nature so much, you should read Darwin's *Origin of Species.*[4] We did not have a copy at home, so I went by bus into Kendal to the public library. I was a bit disappointed initially, because the first chapters were all about breeding pigeons. However, my mother took me to see a pigeon breeder in Arnside who bred fantail, pouter, and homing pigeons that, as Darwin reasoned, had all originally descended from the wild Rock Dove. Some of this man's Arnside homing pigeons had been used to carry messages to the troops in the front line during the war—a practice, he explained, that Julius Caesar had used to send messages to his troops. After this I found Darwin's book entrancing, although still difficult; many details became clearer only when I reread it at an older age, but it did reawaken a question we frequently had discussed as children around the table: Why are we all so different?

Darwin pointed out that variation among individuals of the same species is fundamental to evolutionary change. As a young child, I was acutely aware of differences between individuals of the same species, without knowing that it was vitally important for change over time. For example, my two brothers and I were totally different. John had almost black hair and deep brown eyes; I had dark brown hair and green eyes; and Andrew had fair hair and blue eyes. Not only did we differ physically, our personalities, interests, and future careers also differed (as described in Appendix A). Given these differences, we had fun in the family guessing what another sibling could possibly be like!

Jerry had already shown me how different exposures to sunlight and different soils produced different-looking adult plants of the same species. This clearly was not the reason John, Andrew, and I were so different. John and I had grown up with identical backgrounds—same food, same environment, same parents—while Andrew, born later, may have had different, postwar food. My father used our differences to explain Mendelian inheritance as a fundamental basis of variation, adding that no two people were the same, and he had to take individual variation into account when prescribing medication.

Where our environments did differ was in schooling. The contrast between the level of education and career prospects of boys and girls was striking. The boys' education was at a much higher level academically, whereas at my school in those days, it seemed as though girls were being groomed to become wives of professional husbands. Fortunately, Saint George's School is now a very different place, and academic subjects are highly regarded (as I learned from a search on the Web).[5]

I was not a stellar student. So as not to spoil our holidays, my mother would hide John's and my report cards, producing them only the day before we returned to school. On that day, we would be taken into the surgery one at a time, and my father would go through them in detail. "Och, Rosemary, what is the meaning of this?" he would say. "For every subject except geography there are only four words, 'good on the whole.'" "Oh, I don't know," I replied, "probably average." Clearly, he did not want an average daughter! In those days, he believed that men were innately more intelligent than women, and women who married were not expected to work or have a career. Whenever I said anything remotely intelligent, he would tell me I had a mind like a man.

A mind like a man! Then, in the space of a few days, I heard the same from my schoolteachers. Was I turning into a man? I searched for books on hormones in the school library. One had a figure depicting the average amount and variation of testosterone in males and females. The variation in women was considerable. Off I went to the school matron and said, "I think I might be turning into a man." She took one look at me and in her strong Edinburgh accent said, "Och, no, Rosemary, not with breasts like yours!"

My passion was to become a biologist, and for that I needed to go to university. In my ninth and final year at Saint George's, I pleaded to be allowed to take the university entrance examinations known in Scotland as the higher leaving certificate. I was told firmly that a girl with two brothers should not contemplate going to university, that the money should go toward the boys' higher education because, "After all, you will get married, but your brothers must earn enough to support a family." My father had the same argument. Finally, I wore them down by defiant argument. What if I didn't marry, I would need to support myself. Then: I would rather study biology than marry. I was not sure I believed the last statement, but after that I was allowed to try for a place at university. The resistance to women entering higher education was a strong Victorian belief fostered by specialists in gynecology and obstetrics who gave their ostensible "medical evidence" that mental fatigue after puberty risked sterility. This opinion, prevalent at the end of the previous century, was still very much in vogue. "Intelligent women, unspoiled by education, produced eminent sons. The country would benefit far more from such men than from . . . sterile but educated women."[6]

The time for the Scottish higher examinations came only once a year. Unfortunately, a few days before they were scheduled, I developed mumps with the complication of severe pancreatitis. When I woke up from a morphine-administered fog, there was my headmistress, standing at the end of my bed, hands firmly gripping the yellow iron bed frame. "Rosemary, it is God's will," she said, "you should not go to university." I was not the brightest student in the class, just "good on the whole," having daydreamed through many classes and neglected my homework, but there and then I determined to prove her wrong.

I left school, got a job, and took a correspondence course. This turned out to be the best possible decision I could have made. The correspondence course was so much more advanced than my school courses, apart from geography, and I learned how to work independently. The next problem was how to take the exams. They were always taken in a school. I found overseas students had the same problem and that there was a small hostel in Edinburgh near a hall where the entrance examinations were administered. It turned out the hostel was for men. Nevertheless, the hostel accepted me, and I found I was with a group of men from Africa. They could not have been kinder and treated me like a sister. There was only one bathroom with a series of shower stalls, so they worked out that when they had finished showering, they would knock on my door as a signal that I could have my shower with no fear of interruption. Each evening after dinner, we had fun playing table tennis before returning to our rooms to study. In the mornings, we walked together to Chambers Street, where the examinations were held.

This sideways step worked. I was accepted by the University of Edinburgh, my top choice. The reason it was my favored university was that I considered the Institute of Animal Genetics, headed by Conrad Waddington, one of the best genetics departments in the world—and even better, Charlotte Auerbach and Douglas Falconer would be my professors. Money was a problem. I could not apply for a scholarship, as I had left school. My father kindly agreed to pay for at least one year and perhaps more, if I did well.

PART II

Youth

8

University

> Education is the kindling of a flame, not the filling of a vessel.
>
> —ATTRIBUTED TO SOCRATES

> Education is simply the soul of society as it passes from one generation to another.
>
> G. K. CHESTERTON

The threat of my father not paying for additional years at university hung over me. If I had only one year, I had to make the most of it. I could understand his hesitation; the data were there. Only 3.4 percent of school graduates in the United Kingdom went to university in the 1950s, and among them, the ratio of women to men was one to eleven. Gender roles were clearly defined: men were the head of the household, and women raised children and kept the house. Many businesses, including those run by my uncles, would not employ married women. "After all," they said, "you train them, then they immediately get married and become pregnant." Men were not alone in holding this opinion. Mothers of my boyfriends did not hesitate to tell me they disapproved of the relationship going any further if

I intended to work. "Can you sew?" one wanted to know. "Can you make gravy?" two others asked. "Of course," I told them. Even my mother wrote to her sister saying, "I am afraid Rosemary may be turning into a bluestocking," using a derogatory Victorian term for an intellectual woman not worthy of marriage. My mother was ambivalent. She told me confidentially that when she was younger, she wished to train as a medical doctor, and she frequently discussed with me the plight of women left with no income and no training when their marriage failed. These women were forced either to live out miserable married lives or to raise their children and support themselves on a marginal income. I decided then and there, I wanted to earn enough to be independent, whether I married or not; and surely raising children was the joint responsibility of both parents. My teenage brothers were no help. "No wife of mine will work," they both said, alluding to the social stigma of a working woman. My father was caught in the middle and suggested I apply to the domestic science school in Edinburgh. I could not think of anything worse. I had all the experience I ever needed from helping my mother clean the house, do the washing, iron innumerable shirts belonging to my brothers and father, and cook the meals. This was only training to serve a man, I told him.

Britain was on the cusp of change, and I was just a little early for it. When I applied to universities, there were not many more than a dozen in the British Isles. A few new universities were added in the 1960s, and in 1992, technical colleges gained university status. This dramatically boosted the number of women entering higher education.

All science students were interviewed on arrival at Edinburgh by the dean, Dr. Friend, who lived up to his name. He asked me kindly whether I wanted to become a university professor. I was stunned, embarrassed, and stammered that I did not know if I could aspire to such an exalted height. I just wanted to follow my interest in biology in greater depth.

The first-year lectures in zoology were delivered by senior professors in the department, each one taking three to four weeks throughout the year. We had no textbooks but were required to read primary sources, three or four articles being given to us to read each week. There were a few publications we could buy—one was the book *Animals Without Backbones,* by Ralph Buchsbaum—and another, a small, then unpublished pamphlet written by Charlotte Auerbach called *Notes for Introductory Courses in Genetics,* was given to us.[1]

I was always conscious that I had to work extra hard to keep up with my fellow male students, who, I believed, were innately more intelligent than me. I copied out my scribbled lecture notes each night into a neater form. The joy of doing this was that I could contemplate the lecture slowly, form my own questions, and follow these up in the zoological library the next day. Sometimes my professors would find me there, ask what I was doing, and point me in the direction of papers or books to expand my quest. In this way, I delved deeper into numerous topics such as bird migration, why cancer cells are not sticky, neural regeneration, and embryonic development. I have a vivid memory of sitting in a sunny window in the zoology library one day and thinking, I would love to spend the rest of my life doing this.

I quickly gained like-minded friends, and we had long discussions over coffee and walked in the Pentland Hills watching hares, which turned white in winter. I went to concerts, seated in the Gods in the Usher Hall, played squash, and learned how to ski in the Cairngorms with the university ski club. My friends came from all over the world, some from Africa, others from the Middle East, and several from Iceland. Some of my Icelandic friends had never been to school but had been educated in a system in which the children circulated among several families, spending weeks with one family learning mathematics before moving to another knowledgeable in languages, and so on through music to geology. Most of them were joining the medical school, there being no medical training in Reykjavík at that time. Having been brought up in Viking-colonized Westmorland, I formed an instant bond with them. I found almost no discrimination between the sexes by my professors, but it was striking that men with daughters were by far the most encouraging and supportive—another sign that change was in the air.

Finding somewhere to live in Edinburgh was a problem. While I was on a waiting list for university housing, I found "digs" in a house two miles from the university that took in women students. It was run by a pretentious woman and her daughter. They turned out to be patronizing and racist, strongly disapproving of my African friends. Fortunately, a room soon became available in a women's residence at the university, and I was able to leave. I made friends from all departments but mostly on the arts side.

Initially, I thought I might concentrate on botany and plant genetics, but within the first week I transferred my affiliation to zoology and genetics. The botany lectures and labs were focused on taxonomic classification, whereas zoology was crammed with interesting and exciting questions. At the time of my first-year examinations, I was advised by an older fellow female student not to risk discrimination by putting my full name on an exam paper. I followed her advice, put only my initials before my surname, and did very well in the exams. Continued payment by my father was assured.

In my second year, I took courses on physiology, biochemistry, and histology in the medical school. There I learned about biochemical pathways, gaining knowledge that was of immense value when I did genetics the following year. Teaching followed the tradition of J. B. S. Haldane, in that all our experiments were done on ourselves. One week we tested our blood and urine before and after eating only protein, then repeated it another week on a diet restricted to carbohydrates. One day, I rushed into class slightly late, still in my ski clothes, and the professor pounced. "Ah!" he cried. "Here is a fit guinea pig." I had to run up and down stairs from the bottom to the top, four flights, several times, and then he performed several tests, taking my blood pressure, drawing blood for examination, and so on. After one test I almost passed out. A friend said, "You really worried me, but you did look incredibly beautiful in that state." "Was that the moribund state you would always like to see me in?" I teased.

I became a member of the Biological Society. This was an outgrowth of the Plinian Society, a club founded in 1823 for students only, professors barred from attendance. I was secretary, the same position Darwin had held while a student in Edinburgh. Members of the society met in the evenings about once a month to discuss topics such as the generality of Lorenzian imprinting and the new Watson and Crick DNA model.

One late-summer vacation, all fifteen of us from the Plinian Society went on a two-week camping trip to the island of Jura in the Inner Hebrides. We were allowed to camp in a farmer's field in exchange for a couple of days of haymaking, involving the incredibly hard work of raking the grass into haystacks. It was a magical time, walking for miles through bogs, over heather moors, and through bracken and woodlands to emerge onto isolated sandy beaches, always with the three rounded mountains, the Paps of Jura, visible from every angle. Red Deer abounded. Many students were avid bird-watchers and were rewarded with magnificent views of Ringed and Gray Plovers, Hen Harriers, Merlins, Peregrine Falcons, ravens, choughs, Whinchats, and wheatears. On the coast, we found otters and Gray Seals, as well as whimbrels, shags, guillemots, and my favorite, the Great Northern Diver (or Common Loon). Nearby was Barnhill, the house where, in 1946–47, George Orwell wrote his famous book *1984*. The house was empty, its door wide open, and it was filthy inside. Nearby was the famous Corryvreckan whirlpool. Welcoming a rest after hours of walking, we sat on the cliffs waiting for the turn of the tide and were rewarded by the sight of ferociously churning water, making it easy to conjure up images of sailors being sucked to their death, as is told in tales.

In year three came my chance to do genetics, the very reason I chose Edinburgh. Conrad Waddington and Douglas Falconer, both famous geneticists at the time, were interested in the genetics of complex traits, an area now known as quantitative genetics. It applies to phenotypic traits of plants and animals that are determined by many genes interacting with their environment. Waddington and Falconer were conscious that most students coming to the department from Britain and abroad had very little knowledge and understanding of quantitative genetics, interpreting everything in terms of simple Mendelian inheritance of one gene for one trait. Together, they established a postgraduate diploma course in animal genetics for twelve students: six from the United Kingdom and six from other countries. I applied, not thinking for a moment I would be accepted, and was lucky enough to get an interview. I remember their only question: "Why do you want to do this course, which will take one year of intensive study?" My not very confident reply was that I was interested in the evolution of animals and plants and thought that a knowledge of genetics was fundamental to this question. Amazingly, I was accepted. Everything I learned that year laid the foundation for my future career in biology. It was pivotal to all research I have done ever since.

Falconer was writing his pathbreaking book *Introduction to Quantitative Genetics*.[2] His lectures were one chapter a week from his book, followed by much discussion on each one. All the concepts were new to me, stimulating and mind-bending. As important as attending lectures, discussions, and laboratories was the compulsory attendance at morning coffee and afternoon tea, where everyone gathered, and conversation flowed from one topic to another. We students remained silent most of the time but occasionally were called on for our opinions. Waddington's department was a magnet for well-known geneticists from around the world. Frequent visitors were Theodosius Dobzhansky, memorable not only for his brilliance in integrating genetics with evolutionary ideas using wild *Drosophila* (fruit flies) but also his strong Russian accent; Sewall Wright, who discussed his theory that evolution can occur quickly when populations become isolated; and Michael Lerner, who was interested in the interactions between genes and the environment. They gave lectures, which we all attended, and joined in coffee- and tea-time discussions. The library in the Institute of Animal Genetics was divided into two parts, one stacked full of genetics reprints, books, and journals, the other half with art and literary books. Waddington warned us, "I expect you to spend equal time in each!" His paintings, impressionistic views of chromosomes, hung along the walls in the art section. When examination time came, his warning resurfaced.

The first question was a quote from Voltaire's *Candide* on Gottfried Wilhelm Leibniz's philosophical optimism: *"This is the best of all possible worlds." Discuss this from a geneticist's point of view. You have one hour.* The question was clever because it could be answered superficially or in great depth. I had read *Candide,* fortunately, and I remember enjoying writing this essay, but I fear I would be embarrassed to read it today.

One day Conrad Waddington walked into our laboratory, smoking the pipe that was forever between his teeth, just as another student and I were wrapping up an experiment on *Drosophila*. I asked him what genetic or evolutionary research was being done on animals or plants in the wild. He took the pipe from his mouth, and we had a long, animated discussion about the value of studying populations in their natural environment, and he suggested that the department really should have a first-rate ecologist. He added that he had used wild-caught flies for his heat-shock experiment because he knew laboratory *Drosophila* were depauperate in genetic variation after many generations of inbreeding.[3]

I was an honors student in my fourth and final year. Classes consisting of in-depth discussions on a variety of topics were combined with an independent research project. I discovered that Kay Adams, the only woman professor in the Zoology Department, was working on soil amoebae, fast-growing but wild. She agreed to supervise my study of an interesting problem: eleven strains of soil amoebas, collected from soils around the world, looked similar but had slight differences in dividing time and responses to temperature, and one was suspected of being pathogenic to mammals, having been found in corpses of rabbits. They were hard to tell apart by just looking at them because, being amoeba, their shapes were constantly changing. However, Geoffrey Beale had recently used a serological technique to discriminate between strains of *Paramecium*,[4] and we thought the same technique might be used with amoebas. With the help of a PhD student, Ian Sinclair, who taught me the skill of inoculating and bleeding rabbits without harming them, I cultured the amoebas and obtained rabbit antibodies to them. The results revealed that all eleven strains tested fell into the group known as *Acanthamoeba*. They shared antigens, but no two strains were identical. The one taken from an ill wild hare and thought to be pathogenic was clearly the most different from all the others. Furthermore, differences in environmental factors that I tested, such as temperature and growth medium, had no effect on the results. This proved a nice if somewhat coarse way of genetically characterizing the similarities and differences among the strains and their evolutionary relationships.[5]

To my parents' delight, I graduated honors *cum laude* and was one of a few entitled to wear a hood with white fur trim at the ceremony. The degree on my diploma was second-class honors, but it was explained that this was the highest the Zoology Department in Edinburgh gave—it had not given a first-class honor for the past thirty years! I have always wondered who it was who received that honor, but no one seemed to know.

Spurred on by the result from my amoeba experiments, I began thinking of doing a PhD and wrote a proposal involving Arctic Charr (*Salvelinus alpinus*) in Icelandic fjords. Like salmon, charr spawn in fresh water and once mature migrate to the sea. However, in Iceland, many had become landlocked when fjords were sealed off from the sea and became lakes, and in isolation the fish had evolved into distinct types. In some lakes there were as many as four types: a large and a small limnetic (open-water) form, and a large and a small benthic (bottom-living) form. I was curious to know the genetic and environmental factors responsible for producing these differences and to study the relationships among the fish in the different lakes. I thought these changes must have occurred quite recently, after the last glacial maximum, which ended around 12,500 years ago. I planned to determine when the fish had been isolated and to study the ecological differences between fjords in order to understand why the fish had changed in morphology and behavior. In addition, I intended to collect fish in the field and rear them in the laboratory for behavioral tests. I also wondered whether the serological technique I had used in amoebas could be used to find the genetic relationships among types and populations from different fjords. If so, this could give a genetic underpinning to the morphological differences.

My dream project became even more exciting after discussions with my closest friend at the time, David Kinsman, a geologist. He planned to go to Iceland and told me that it was quite likely he would be able to determine the precise time at which the charr in many of the fjords had become landlocked, based on the timing of volcanism that caused a barrier between the fjords and their exit to the sea. I was all set to go, once again very much against my parents' wishes. This was also against the wishes of my professors, who liked my project but were worried about me carrying out my research alone, as a single woman. However, in the meantime, Professor Michael Swann, the head of the Zoology Department, had received a letter from Professor Ian McTaggart Cowan, the head of Zoology at the University of British Columbia, asking him to recommend someone to teach the third-year embryology labs for a year. He recommended me and told me that I would gain valuable teaching experience, earn some money, though not much, and could return to Edinburgh to do my PhD research on charr. This was a dilemma, Iceland or Canada? The University of British Columbia and the mountains beckoned, and I decided on Canada, with the option of returning to Edinburgh for my PhD. When Professor Swann saw my letter of acceptance from Professor Cowan with the proposed salary, he exploded: "You cannot possibly live on that! Write to him and say you will only come if they offer you more, but do so politely!" I did, they did, and I went.

9

Vancouver

If you can look into the seeds of time,
And say which grain will grow and which will not,
Speak then to me, who neither beg nor fear
Your favours nor your hate

—WILLIAM SHAKESPEARE, *MACBETH*

Canada: a new country, a new life, and my first serious job. I had intended to continue at Edinburgh, doing my PhD research in Iceland, but this unanticipated opportunity was too good to miss, and I thought it would interrupt my plans for only one year. Just one step sideways and back again. How wrong I was!

My father gave me two articles that he said I must read. Both were extensions of our many discussions at home about the impact British colonization had had on the world and were so different from my school history books' glorification of the British Empire. The first article reported that over 60 percent of the Canadian Indigenous population had died during the 1800s from smallpox, typhoid, measles, and syphilis, all introduced from Europe and previously unknown among the peoples of the Americas. The second was on what my father considered to be an insidious residential boarding-school system that isolated Indigenous children from their own parents, culture, and religion, with the intention of assimilating them into the mainstream Canadian way of life. Accordingly, I entered Canada not only excited about my first teaching job and the opportunity to explore the relatively untouched wilderness of British Columbia but with a desire to gain some knowledge about the cultures of the First Nations peoples.

Getting to Vancouver with little money in 1960 was a challenge. The cheapest way was to travel in steerage class on one of the ships sailing to Montreal and then to fly from there to Vancouver in small hops on small planes. After a rough, seven-day sea crossing, the weather changed, and we glided into the Saint Lawrence Seaway. Leaning on the rail, I caught a glimpse of a few whales in the distance, and then we began our long, slow passage down to Montreal, passing small towns with white wooden churches standing out against a coniferous forest background. Instead of the salty winds off the ocean, the breezes coming from the late summer land had a spicy, dusty tang. In Montreal, after the usual hassle at immigration and customs, I went to the airline desk and found the main airlines were on strike. I felt lost and alone but eventually, by chance, stumbled across some airline offices near a large central hotel and managed to exchange the first of my tickets for another and begin my hops across Canada that night.

I arrived in Winnipeg at two a.m., hoping to exchange my other ticket. I was helped by an old farmer who had been sitting next to me and had regaled me with stories of how the wonderful invention of the combine harvester had changed his life. He helped me to get my next flight, to Edmonton, and I strongly suspected that he kindly had paid the difference between my old ticket and the one he obtained for me, although he shrugged it off when I offered to pay.

This plane was small and low-flying. At dawn, the land beneath us appeared like a tapestry: fields of wheat interspersed by those left fallow, threadlike roads connecting farms, each surrounded by a windbreak of trees, often with a pond nearby. An occasional river appeared, meandering snakelike across the landscape. The ever-present railroad ran straight as an arrow, east to west, punctuated by regularly spaced grain elevators. I recalled my old question in Arnside: What was this like before people arrived? I conjured up visions of American Bison roaming the prairies and Grizzly Bears coming down from the Rocky Mountains in spring with their young.

It was snowing when we arrived in Edmonton early on September 3, 1960. How far north were we? I wondered. There I undertook yet more negotiations for another flight, this one to Vancouver. I was longing to fly low over the Rocky Mountains, but I had not slept since the last night on the boat, and the next thing I knew, the flight attendant was waking me up for the landing in Vancouver. I had missed not only the mountains but also my only meal. Amazingly, when I arrived in Vancouver, someone was there to meet me. She introduced herself as Mary Jackson. How did she know I was arriving? Somehow Professor Cowan had found out I was on that plane and had sent her in her car to meet me. She also explained that she was looking for someone to share her apartment. I was not only met but also given an instant home, and only a few days later, Mary introduced me to the magnificent West Coast wilderness.

Mary and George, her boyfriend, had planned to climb Mount Baker the weekend after I arrived and asked me if I would like to join them. I explained that I had never used crampons or ropes. "Never mind," they said. "We will teach you." On our first night we camped just below the tree line in old-growth forest of tall, lichen-covered Douglas-fir, Mountain Hemlock, and Western Red-cedar. The next morning, we emerged from the forest into an alpine meadow of heather and bearberry shrubs and walked toward a glacier, where we donned crampons and tied ourselves together with ropes. The glacial crevasses, of which there were several, were terrifyingly beautiful. I peered down into their depths, where their icy walls reflected blue and green light. I was relieved to get to the other side, where small plants and warm cushions of moss hugged the stones. We lay down on them exhausted. From there the climb was easy, but we still had to repeat the glacier crossing on the return. I could not help wondering, if one of us had slipped, would the other two have had the strength to hold him or her? Mary was light, but George looked rather chunky and weighty to me. I decided that hiking, not climbing, was the life for me.

The next weekend, Mary, George, and I went to the Olympic Peninsula, which had been a glacial refugium, home to several endemic plants and animals, including a marmot, chipmunk, and elk. Glacial refugia are like isolated islands, or like the Icelandic fjord-lakes, where organisms, like the charr, become trapped and evolve along diverse evolutionary pathways in isolation. I was excited to see the features that distinguished the peninsula's marmot, chipmunk, and elk from their close relatives—although the differences seemed to be rather small, from Mary's descriptions. The first afternoon we spent at the foot of the mountain to the northwest. This was my first experience of a temperate rain forest. Enormous thousand-year-old Western Red-cedar, Sitka Spruce, Douglas-fir, Bigleaf Maple, and Western Hemlock trees towered above us, all draped in epiphytic lichens. At our feet were fallen trunks, yards deep, covered in mosses and ferns, slowly rotting in the warm, wet atmosphere, recycling nutrients back to their living offspring. After exploring the rain forest, we decided to camp higher up so we could hike in the alpine meadows the following day. I will always remember the sight of those meadows under clear blue skies the next morning. Grasses, sedges, and the remains of summer flowers covered the thin, stony alpine soil. Crystal-clear lakes reflecting green and blue light were surrounded by rocky cliffs with fern-covered ledges. Here we watched rosy finches and heard ravens call, that deep, hollow croak that carries for miles. We found the Olympic Marmot (*Marmota olympus*), which was slightly larger than other marmot species and, according to Mary, had different calls. The Olympic Chipmunk (*Neotamias amoenus caurinus*) has black-and-white stripes, and the Olympic Elk (*Cervus canadensis roosevelti*) is larger than other members of *Cervus canadensis*. As we were leaving, a herd of these huge Olympic Elk rushed in a panic down a scree slope in front of us.

When we returned to Vancouver, Mary asked me if I would like to meet a fellow English person. I was not especially thrilled to meet an Englishman. I was enjoying the freedom of being in a country where the first questions on being introduced were not "Which high school did you attend?" and "What is your father's occupation?" I was relishing being my own person in a new country.

I had experienced some of the British Columbia wilderness, started my teaching career, and managed to acquire a somewhat livable wage. But as it turned out, by far the most momentous experience was meeting Peter, a completely different fellow English person from the one I had envisaged. He tells quite a different story of that first encounter, in which I wheedled a secondhand pair of skis from him.[1] All true, but I have conveniently forgotten that part. My memory is more embarrassing. A few days after we met, Peter and I had coffee together and then walked back to our respective homes along University Avenue. I stopped at a drugstore to get some office supplies. He wandered in with me. I found my pencils but could not find a rubber, so I asked the assistant, and she pointed me to the pharmacy. Why the pharmacy? Peter fled, and she blushed. Why? When I joined Peter outside, he was doubled up with laughter and said, "Do you know what a rubber is in Canada?" I did not. "You wanted an eraser," he said, "and instead you asked for a contraceptive for a man, and I was standing right beside you!"

Peter was going to Mexico to do fieldwork for his PhD research on birds. There was an Olmec exhibition in the Vancouver Art Gallery, and he asked me if I would like to go with him to view it, after which we would have dinner at an elegant Mexican restaurant. Now it was Peter's turn to be embarrassed. After a visit to the museum and a beautiful Mexican dinner, with wine, he discovered and then confessed that he did not have enough money to pay. In those days, it was always the man who paid, especially on a first date in a chic restaurant such as this, so when we were sure the waiter was not looking, I slipped the money under the table into Peter's waiting hand. It took us an hour to walk home. There was no money left for bus fares!

Having recovered from our blunders, we began to enjoy mountain excursions together. We bought sealskins to attach to the bottoms of our skis so that we could climb into the snow-filled alpine meadows. Diamond Head was our favorite destination. After a long trek up the mountain through a snow-covered Douglas-fir forest, we emerged into an alpine meadow with the speck of a small cabin in the distance. As we approached it, the aroma of freshly baked bread drew us magnetically forward. Millie, a Norwegian and a friendly tyrant, was the baker and owner of the hut. She greeted us with the words, "If you are not married, there is no sharing the same bedroom." A marten peeped out from between the log cabin's double walls where he had made his winter home, as if ensuring the rules were obeyed. On these trips, we usually skied on the wide-open snowy slopes all day and then at dusk raced each other down toward the snow line, often tumbling over as we suddenly encountered the gravel road.

George, Mary's boyfriend, gave me my first glimpses into the region's Indigenous cultures. George was a stocky, somewhat taciturn man with his head permanently at a sideways tilt. His background was vaguely mysterious; he never referred to a family but seemed to have somehow escaped from Czechoslovakia as a boy and arrived alone in Canada, perhaps during the coup d'état in February 1948, when the Communist Party of Czechoslovakia seized power with the support of the Soviet Union. I never knew his age. In Canada, he had trained as a surgeon, but sometime shortly after he arrived or later, he spent two years living with a group of Inuit people. He was a fount of knowledge about their culture and enormously impressed by their ability to live in harmony with their environment.

Peter and I saw something of George's Inuit-acquired prowess when we spent a weekend skiing with him and Mary in the wilds of Manning Park. George chose the place to make a four-person igloo in deep snow near a frozen lake. He taught us exactly how to make it, complete with an entrance tunnel to crawl through and a slab of ice, perfectly cut with a machete, inserted into the wall for a window. That night we slept in a row, careful to blow out the candle to prevent the snow melting inside and dripping down onto us. When we crawled out the next morning, George was already cooking breakfast over a log fire, with one hand brushing away Whisky-jacks (Canada Jays) that were intent on stealing bits of egg and bacon from the frying pan. We spent the remainder of the day skiing in blinding, blowing powder snow and then sadly returned home in George's ancient Volvo.

FIGURE 4. Upper: Flatiron Mountain, British Columbia, where Peter and I became engaged. Lower left: Peter, in the photograph I sent to my parents informing them of our plan to marry. Lower right: Totem pole on University of British Columbia campus.

Peter and I visited the igloo six weeks later, on our way to the Rockies, and it was still there, in perfect livable condition.

Bristol Foster was the second person I met who had a deep knowledge of Indigenous cultures; this time it was the Haida, First Nations people who live on the Haida Gwaii archipelago. Peter shared a room in the Department of Zoology with Bristol, a tough, tall, lanky Canadian PhD student. The room was also home to Bristol's Short-eared Owl (*Asio flammeus*), known to everyone as Howland. Bristol fed Howland each night by throwing dead, thawed mice down the length of the corridor to make them appear alive. It fooled Howland; he caught each one! Bristol prided himself on his wine-making skills, and several flagons were bubbling and occasionally exploding under a bench. He had just returned from a year driving across sub-Saharan Africa with his artist friend Robert (Bob) Bateman. Bob had decorated the yellow Land Rover that Bristol was still driving with eye-catching paintings of wildlife from this trip. Bristol was doing PhD research on Haida Gwaii (also known as the Queen Charlotte Islands), comparing the small deer there with the larger mainland ones of the same species and the large deer mice there with the smaller ones on the mainland. He concluded that large mammals get smaller when food supply is limited, and small mammals get larger in the absence of predators.[2] While on the islands, Bristol became immersed in Haida culture and met Iljuwas Bill Reid, who became his lifelong friend.

Peter and I frequently encountered and talked with Iljuwas Bill Reid on our lunchtime walks through the University of British Columbia forest in 1960–61. He was forty years old and in the process of restoring intricately carved totem poles that had been abandoned on Haida Gwaii for the reconstruction of a Haida village that would later become incorporated into the UBC Museum of Anthropology. Reid's father was a Canadian of Scottish German descent, and his mother was Haida but had been denied her native heritage under the terms of the Canadian government's oppressive Indian Act of the time. Bill Reid became a CBC radio announcer in Toronto, and it was not until he was in his twenties that he returned to Haida Gwaii and met his grandparents and great-great-uncle, who were Haida silversmiths. Bill Reid's artwork is now known throughout the world. His *Spirit of Haida Gwaii: The Jade Canoe* is the focal point of the international area in Vancouver Airport, and *Skaana—Killer Whale, Chief of the Undersea World,* a breaching Orca, is in the Vancouver Aquarium. These are massive structures that tell stories of Haida myths. Just as impressive is his intricate and beautiful jewelry, inspired by the work of his maternal grandfather and great-grandfather. Many pieces are now in the Museum of Anthropology at the university. Reid spent the rest of his life fighting for the respect and knowledge of his culture, and as *Iljuwas,* he magnificently did his part to change the world.[3] He was one of the most inspirational, influential and admirable people I have ever met.

My job, teaching embryology in the laboratory to third-year students, started in mid-September 1960. The student body differed strikingly from the equivalent in Britain. There were as many women as men in the class, and many of the students were working to put themselves through university; consequently, some were several years older than I was. It seemed as if every one of them had a unique story to tell, very different from the traditional progression from high school to university that was typical in Britain, except for a two-year interruption for men doing their compulsory national military service. A few of the students were Japanese Canadians who, as children with their parents, had been forcibly interned in government camps in the interior of British Columbia during the war. But conspicuously, there were no First Nations students, and I wondered why not. For teaching the class, I relied heavily on my genetics labs in Edinburgh and Waddington's textbook *Principles of Embryology* for planning and instruction.[4] I think I would be embarrassed today to see how I taught, having learned so much since that time, but I hope youth and enthusiasm compensated in part for lack of skill.

I soon found that my salary was meager, and after rent and food had been paid for, there was not much left. I managed to get a second job as a research assistant to Dr. Cyril Finnegan, who needed someone to make hanging drop preparations of cells and stain them for RNA as part of his research on the development of amphibian embryos. This I could easily do, and I got quite adept at doing many at a time. By holding the two jobs, I exactly doubled my salary.

Peter and I were spending an increasing amount of our time together. Our regular weekend skiing haunts were two local mountains, Grouse and Hollyburn. These were easily accessible by car with friends or by bus from the university, followed by single open chairlifts to the top. In March 1961, we decided to go farther afield and traveled by train and then on foot to Whistler, there being no road at that time. Zigzagging up through forest trails, we eventually reached the snow, and then climbed farther yet for another couple of hours in a white landscape with nothing but a single deserted, broken-down log cabin. Small jumping insects known as snow scorpionflies (*Boreus elegans*) were feeding on red algae encrusting the snow. Knowing that Geoff Scudder, an entomologist at UBC, wanted some, we collected several in the empty vials Peter always seemed to have in his pocket, along with spare string and years-old toffees that he said were kept for emergencies. Today, this once wild open space is now a world-class ski resort for the fashionable, even hosting Winter Olympics competitions in 2010, accessible by a major highway from Vancouver.

Peter's background was very different from my own, which had been stimulating and made secure by loving parents. Peter's parents divorced when he was two years old, and custody was given to his father. He spent his third year with his father's sister, who had a son a year older than Peter. At the age of three, Peter was evacuated from London during the war to a boarding school, where he remained until he was eight years old. It was hard for me to imagine a mother leaving her child. What kind of person was I getting to know, this man who had mostly blanked out from his memory the formative years from three to nine and who had several divorces in the family? When we started to become serious about the possibility of spending the rest of our lives together, we had long and earnest discussions. I remember one time when we sat down by the shore and talked for hours about our ideas of sharing and, importantly, our philosophies on raising children. We both thought that giving children the best education possible was the next highest priority after the top priority of providing a loving, caring, and stable home environment for them. I was amazed that Peter, in view of his background, was so well grounded, although I was not surprised that he was so independent. My mother had gone through the same memory amnesia after her mother died when she was eleven and she was sent to a boarding school until she was eighteen without breaks for vacations. She could not remember or would not talk about those adolescent years. Peter and she were alike in many respects, including being more routine-bound than I was, as though routine were a way to impose stability in their lives.

It was exciting to find that Peter and I had a joint interest in the causes and processes of speciation. Peter approached the subject from an ecological point of view and was especially interested in the importance of competition for resources. He was about to embark on a PhD project in Mexico that involved comparing populations of several species of birds on the Tres Marías islands with the same species on the adjacent mainland to understand their differences in size, song, and color and their relevance to species formation. I agreed these were important, but my perspective was different. Drawing upon my genetics background and thinking back to glacial refugia and my pet project on Icelandic charr, I thought that sufficient genetic variation was also vitally important, as it determined what evolutionary change was possible. My questions, then, were: How are measurable (phenotypic) variation and genetic variation maintained in a changing environment, and how are two species formed from one? Our two approaches to the same question of the processes involved in speciation were complementary. So began our long explorations and exciting journey through life together.

10

Mexico

> Learn how to see. Realize that everything connects to everything else.
>
> —ATTRIBUTED TO LEONARDO DA VINCI

The prospect of visiting Mexico was thrilling. We both devoured books and articles related to the plants, animals, and cultures of Mexico and South America. I found it astounding that at the height of the ancient Greek civilization, in 500–400 BCE, halfway round the world in Mexico and Central and South America, peoples unknown to Europeans not only had been cultivating crops such as maize, squashes, beans, and peppers for thousands of years but by 700 BCE had built cities dominated by massive temples that were decorated with elaborate facades of their gods. Early American civilizations' religions were polytheistic, like that of the Greeks. In 500 BCE, these peoples had developed an accurate calendar. By the third century BCE, they were using hieroglyphic writing to record their history, and sometime around 100 BCE they had developed the mathematical concept of zero. Like all human societies, these ones had complex relationships with neighboring cities and clans, involving at times peaceful trading and at other times, when resources were scarce, wars and rituals involving human sacrifice. The highlight of all this reading was finding a translation of Alexander von Humboldt's diaries in the rare-book department in the basement of the library at the University of British Columbia. Humboldt's impressions and descriptions of the geology, plants, and animals along with the local cultures of South America were riveting. He was horrified at the extent of the colonial Spanish exploitation of the Indigenous peoples in the conquistadors' greed for gold and silver and their treatment of them as slaves. He held a strong conviction that all peoples were equal and that humankind was part of the whole diversity of nature and should be free to live in harmony with nature. After leaving Mexico and talking to President Thomas Jefferson in Washington, DC, Humboldt reiterated his concern about slavery and exploitation of people, knowing full well that Jefferson himself enslaved people.

Peter left early to start his research. As soon as I finished teaching, I flew to Mazatlán to join him briefly before returning to England. Peter met me at the airport, and together we walked through the shimmering heat down to the coast. For me, it was sensory overload. First vivid impressions were smells of overripe melons and papayas, squadrons of pelicans flying in formation over an incredibly blue sea, their beaks thrust forward like bayonets, frigatebirds circling overhead. It was late afternoon, and the sun sank fast over the horizon. No northern twilight here; darkness was instantaneous, triggering a chorus of insects and the sudden appearance along the seawall of courting scorpions, engrossed in their intricate dances.

Next day, we flew to Puerto Vallarta in a small three-seater plane. In 1961, Puerto Vallarta was a tiny coastal village with no road, and the only alternative access was by sea. We spent the next few days looking for the birds Peter was studying while climbing up through the dry broadleaf deciduous forest. Another surprise, although obvious when I thought about it—the trees were shedding their leaves in the hot, dry season. I remember *Bursera, Ficus, Tabebuia, Croton,* and *Cordia* trees, but there were many other species, all unfamiliar. *Bursera* is prized for the resin that oozes from the trunk and is used for incense burning in churches. We met few people apart from the occasional chachalaca hunter. Chachalacas are like arboreal turkeys, and I imagine they taste like them. One afternoon, the sky above us suddenly turned dark. At first, I thought it was a solar eclipse, but no, it was a cloud of migrating dragonflies, flying past us for a couple of hours and heading north; an amazing sight that made us wonder why they do it and what had triggered such a spectacle. We know now from scientists who have tracked these movements that a drop in overnight temperatures for two consecutive nights gets them going.[1] Although an individual dragonfly can fly nine hundred miles on its tiny wings, the trip from Mexico to Canada and back is multigenerational, with one generation in Mexico being sedentary, just as with the better-known migrator the Monarch butterfly.

Leaving Mexico in 1961 was a different kind of adventure. Those were the days when girls and women wore long skirts and were always chaperoned by their husband, mother, or an older woman. I boarded the plane alone for the first hop to Mexico City, wearing appropriate attire. The pilot appeared out of the cockpit and, concerned that I was unaccompanied, immediately found four nuns to look after me. When we reached Mexico City, he kept me on the plane and found a taxi driver to take me to my hotel but told me sternly to say *NO* if the driver suggested taking me to see the murals at the university. This was disappointing because I had planned to see the murals by Diego

Rivera and José Clemente Orozco, thinking I would have enough time between flights. I did as I was told, and the next day I was escorted by a different set of four nuns from the hotel to the airport. I caught up with murals by Orozco thirty years later when Nicola, our daughter, was an undergraduate at Dartmouth College.

In Dartmouth's main library (Baker Library), Orozco painted a series of powerful frescoes in 1932–34, at the height of the Great Depression. He named the work *The Epic of American Civilization.* According to the college's brochure, Orozco wrote a press release in 1932 stating that his series "represents man emerging from a heap of destructive machinery symbolizing slavery, automatism, and the converting of a human being into a robot, without brain, heart, or free will, under the control of another machine. Man is now shown in command of his own hands, and he is at last free to shape his own destiny." The two panels that haunt me are centered on Quetzalcoatl, the mythical plumed serpent god. Like the Muses, those beautiful women that inspired creativity in the Lyceum in Athens, Quetzalcoatl is depicted as an inspiration of creativity in humans through collaboration. The mural named *The Pre-Columbian Golden Age* shows the cultivation of crops and references a knowledge of mathematics, writing, and astronomical understanding in the construction of a calendar. The message is peace, creativity, and tolerance. The following murals signal the mayhem that occurs when Quetzalcoatl leaves—wars, greed, and institutionalized schools in a society that destroys creativity and favors the wealthy. The final panel brings hope, with a worker reading a book, the symbolic return to another golden age of tolerance and access to education for all. We would be naive to think that slavery and inequities between rich and poor did not occur during the golden ages of Greece and pre-Columbian Mexico, but the message does inspire us to work toward tolerance and education for all.

The Orozco murals were commissioned by Dartmouth College president Ernest Hopkins and paid for by Nelson Rockefeller, a Dartmouth alumnus. They were not well received by alumni and parents, to whom Hopkins replied: "As to the message not being nice, if that be a criterion of judgment many of the great works of the medieval masters would have to be removed from the Louvre." This speaks to Hopkins's admirable commitment to tolerance and openness to new and perturbing ideas.

I stepped off the plane in London, no longer escorted by my four nuns, and took the first train to Arnside. Jumping down onto the station platform, less than a year after I had left, felt like stepping back a century. The tangy estuarine smell of thyme and salt and the call of curlews had not changed; only I had changed.

11

Marriage

> Love looks not with the eyes, but with the mind,
> And therefore is winged Cupid painted blind.
>
> —WILLIAM SHAKESPEARE, *A MIDSUMMER NIGHT'S DREAM*

Met in 1960; engaged in 1961; married in 1962. We became engaged on a camping trip many miles above the Ashnola River in the interior of British Columbia on Flatiron Mountain in Bighorn Sheep country, where Peter backed up his proposal of marriage with an engagement ring he made by twisting a *Vaccinium* (blueberry) stem into a neatly fitting circle. When I arrived home in Arnside in the summer of 1961, my father had already received a letter from Peter (prompted by me) officially asking for his blessing of our intended marriage. His tongue-in-cheek reply was that he would be delighted to "give the hand" of his daughter in marriage, but he regretted there would be no dowry as my education had been "expensively neglected"!

That summer, many letters passed between us. Peter wrote frequently from Mexico describing the delights of fieldwork experienced in a new environment but also his complete disregard of danger in pursuing his dreams. What kind of daredevil was I about to spend the rest of my life with? Desperate to get to the Tres Marías islands, he had hired two fishermen with a dugout canoe powered by an ancient motorbike engine. A mast and a piece of cloth as a backup sail stowed in the canoe provided their only safety margin. They planned to eat what they found on land and caught in the sea. The port captain was reluctant to let them go—it was hurricane season, after all! Somehow, finally, they persuaded him to allow them to go.

Peter's letter starts with his excitement at reaching the first of the islands, María Cleofas:

Till my last day I shall never forget the beautiful bay on M. Cleofas into which we pulled after the sun had sunk behind the hill. Wonderfully calm, green water in which a turtle was lazily flapping and one Pelican diving, a thin white ribbon of a beach made of coral fragments, land rising to about 20 feet all the way around, clothed thickly in cactus and on one side a superb outline of Magueys. On some of these plants, which grow up to 30 feet, Pelicans were roosting. While we were unloading I suddenly looked up to find the sky a green-grey and the thin wisps of cloud present were vividly scarlet, a wonderful combination of colours.

The next day they reached their true destination, the island of María Magdalena.

I was in a very light-hearted mood when I left the camp and walked over coral fragments along the beach. With the impenetrable tangle of cactus and thorn on my left that had been present on M. Cleofas—very similar to the vegetation on our hill at Mazatlan. I was looking for the Barranca or Arroyo which Dr. Larkin had visited before, and after a quarter of an hour found it. . . . I could pinpoint to the very second and the very spot where I stood when my heart gave a perceptible leap when I saw the opening of the Barranca. It was more than beautiful, it was overwhelming, it made my original pleasure at arriving on the island seem as small as one's reaction of amusement to a joke. . . . And as I walked further and further in I knew my initial expectations were being lived up to. The nearest to this, the most wonderful walk I have taken for a very, very long time I can recount, is the English countryside in a Beech forest. . . . There are no palms. The soil is clay. The leaves are beech brown and lie on the ground I should imagine for several years and hence form a carpet. There is little cactus and on the whole little herbs and shrubs altogether. . . .

Instead the canopy impresses: not dark and overwhelming but . . . allowing the penetration of a fair amount of light. Red-barked trees are present I am glad to say and so is another even stouter and taller hardwood, a truly regal tree and from it—another surprise—a Spanish moss type of lichen. And through all this runs the Barranca—now known as Barranca Rosita a 5–10 yd wide dry river bottom, with sheer 4' or more walls and covered with leaves, a few branches and the occasional fallen tree. After two miles of this I returned elated. Birds were abundant, furthermore they were singing in customary English woodland fashion, the Robin counterparting the Song Thrush, the Red-Eyed Vireo the Robin, the Parula Warbler the Willow Warbler, etc., etc. Both the Tropical Kingbird and the Happy Wren were there too. . . .

Next day! Next day, up early, off to the same Barranca and a long hike up it to a very small waterfall—not functional but with water at its base, on for a further mile and then up to the high ground, hoping that I had selected the highest point on the island. It was relatively easy to move around after leaving the barranca, there being some creepers obstructing one's progress, but not so bad as at Puerto Vallarta by a long way. I had lunch on this peak—not the highest point of the island, but only of the south. Then the trouble started. I started descending in the wrong direction and got lost. Stupid Pedro. I returned to the lunch spot, took my bearings again and set off once more: this time I was not stupid but just unlucky to choose the wrong barranca finishing an hour and a half at 300′ above a sharp drop to the sea! I then played safe, and directed myself at 180° to the setting sun and set off, climbing trees and using deer tracks to help. From this distance it sounds fine but it wasn't: I was simply lost, and resigned to sleeping out somewhere and making camp the following day. Towards sunset, very tired, I deserted my plan and chose to follow a barranca hoping that it would bring me out somewhere on the correct coast, cross-country travel being far more exhausting. . . .

Eventually about an hour after sunset with 1/10 of a foot candle of light in the sky I reached the sea. . . .

Funny, looks familiar. Turn right and walk a little way to see if the campsite is at the far end of the Bay: and this led me to the entrance to Barranca Rosita and back to camp a little later. A very sobering experience from which I hope to learn a lesson that will be well learned. All due to my carelessness and overconfidence, it could easily have been avoided.

Peter and his crew had still to return to the mainland, and now a storm was brewing, so they first went to a better shelter on the neighboring island, María Cleofas:

We left the island just before sunset. Bad luck it seemed had arrived yesterday with my return to camp and was destined not to leave us. A two-hour journey to M. Cleofas took 6, thanks to the damn engine, or rather the gasoline. Utterly exhausted we went to bed well after 1 o'clock and were up again at 5 o'clock. Bad weather ahead. Shall we leave. Well we can only try. An Hours coaxing got the engine started, and an hour and a half later the Captain quite rightly decided to turn back. As we turned back my spirits could not have been lower. 6 hours sleep only in 2 days and frustration left me exhausted. . . .

Early to bed, up even earlier at 3 o'clock, and we left at 5 o'clock on a calm sea. But we were destined to have things our own way for a short time; a head wind got stronger and stronger and by 10 o'clock it was raining. This was the first time I had been goose-pimply in the Tropics. There was no turning back however, we were committed with barely enough gasoline to reach the mainland. . . .

The storm weathered, we almost reached land and then the engine packed up for the last time. So up went the huge sail we were carrying and we arrived less than an hour later. Unlucky in some respects but lucky in others. In mid ocean the sea seemed very cruel, I tried to think how much and where we would have drifted had the engine packed up there. As it was we spent the last night on the Punta, removed a visit to Las Tres Marietas from our list, very regrettably, and sailed home, doing very well to make the 2½ hour journey in 5 hours.

This I enjoyed very much once the wind had been caught. I had never been sailing before, what a way to start! On landing scores of sailors and children buzzed around us, chattering at two hundred and thirty eight to the dozen. There then followed the usual journey to the hotel, drinks in high spirits and muchas gracias.

The Captain and Diego were really excellent company throughout the whole of the trip, doing all the heavy work and allowing me thus all my time on my birds. We did not fall out over anything and remained and parted very good friends.

After this, I added a compass to the emergency supply of toffees, string, and vials in his pocket.

Meanwhile, in faraway Arnside, there was excitement. Villages love a fiesta, and here was an excuse. News of my impending wedding spread, and it quickly became a village affair. Everyone wanted to be involved. A friend of my mother's made my dress. Connie Ashworth, a longtime family friend, made the cake. Mr. Dod, the baker, did the catering and gave us one of his paintings of Arnside shore. I treasure this painting; it hangs on the dining room wall as I am writing, sixty-one years later. The reception was planned to be in the village hall so that everyone in the village could attend. Arrangements were made to start heating the stone church a week before the ceremony, and all the choir boys' surplices that had not been cleaned for many years were sent off to the cleaners. All this was because the villagers wanted to express gratitude to my parents, who at one time or another, through their involvement, concern, and kindness, had helped them. Peter and I were like grateful puppets with the inhabitants of Arnside pulling the strings.

January 4 was a frosty, sunny day, and Arnside was sparkling. At two p.m., exactly, Felix Mendelssohn's "Wedding March" began. I walked with my father slowly down the aisle wearing my recently made wedding dress and a veil that had been passed down through several generations on my mother's side. As I stood parallel to Peter at the altar, he leaned over and whispered, "I heard on the radio this morning that the DNA code has been cracked!"

We returned to Vancouver via New York. It was my first visit there, and I found the city claustrophobic, the tall skyscrapers

FIGURE 5. Left: My parents' wedding, 1935. Right: Our wedding, January 1962. My mother and I are wearing the same veil.

overbearing and blocking light on its way down to the streets below. It was nothing like the wide-open spaces surrounded by beautifully designed buildings of European cities. A freezing January wind whipped and pierced through our inappropriately thin English clothing. Nevertheless, we were in love and had fun. We spent the days measuring specimens of Mexican birds in the American Museum of Natural History, and each evening ate at a different ethnic restaurant. In between, we squeezed in visits to the Metropolitan Museum of Art and the Metropolitan Opera. On our return to Vancouver on Air Canada, the pilot emerged from the cockpit and congratulated us on our marriage. He was one of my students and left saying, "See you in class on Monday!"

Money from my two jobs and Peter's stipend gave us a meager but adequate income. My parents gave us enough money to buy a Volkswagen Beetle. This allowed us to continue camping and exploring the British Columbia wilderness every second weekend. No longer did we have to rely on the train and a Greyhound bus to drop us off at trailheads. George told us that his most difficult operations were on people who had been in VW accidents or mauled by a Grizzly Bear. "Be very careful," he said. "For every bear that you see, there are ten others you haven't but who have seen you!"

Our first home in Vancouver we named "The Burrow." It was a rented basement containing one room, bedroom plus kitchen, next to the furnace room with its tangle of three huge pipes we called the trifid. The bathroom was shared with our landlord upstairs. The floor was black with a huge yellow question mark in the center, inside which was another question mark, this one in red. It was so hideous that, with our small salary, we bought a secondhand rug to hide the gruesome pattern.

Peter's PhD research in mainland Mexico and the Tres Marías islands was producing intriguing results. He was finding that for many of the species, the island birds had larger beaks and longer legs relative to their body size than their mainland counterparts. Diet and song also differed between islands and the mainland. The island birds had fewer predators and larger population sizes than their mainland relatives, and the question was why. Peter explored possible ecological reasons for the differences by comparing the island birds with their relatives on the mainland at Tepic in the highlands and at coastal Puerto Vallarta. He found that those with longer beaks relative to their mainland counterpart consumed a greater range of food items, and those with longer legs used a greater variety of perches.[1] When I was released from teaching, I flew down to become his research assistant, enjoying the challenge of seeing birds I had never encountered before in a completely new environment and the biological questions they raised.

Doing fieldwork together in the forested mountains of Tepic and in Puerto Vallarta was fun and exciting. We had adventures, including one on a drive along burro tracks from Tepic to Puerto Vallarta. Halfway there our car spluttered and stopped on a steep sandy hill. We had not seen a person for at least thirty miles, nightfall was approaching, and we thought we were alone in the forest, when suddenly about a dozen Huichol men appeared and pushed the car over the summit. Then, using signs, they guided us into a large round dwelling made of thatch, where several women were cooking river fish. They gave us fish washed down with boiling-hot goat's milk. Everyone including us went to sleep, darkness having fallen. When we woke the next morning, they had all vanished, and all we could do was to leave a little packet of money as a thank-you. The car must have overheated, because it started now, and we resumed our journey, only to meet a shallow river, the Río Ameca. I waded ahead to test the depth and position of the rocks while Peter followed, driving the car. Only after I got to the other side did I wonder what was in that muddy, slow-moving river. Somehow, we managed to reach our destination, the small hamlet of Puerto Vallarta.

On our first visit to Puerto Vallarta, we woke up each morning to the sounds and smells of tortillas being made; the slap, slap of corn dough being pounded and flattened between hands before it was roasted on stones around an open charcoal fire. On our last visit, only three years later, tortillas were being made on an electric contraption resembling a conveyer belt. Gone were the early morning sounds and comforting aroma, but there were many happy women with reduced chores. Groups of women still regularly washed their clothes in the river, laughing

FIGURE 6. Mexico, 1962. Upper: Río de Cuale, Puerto Vallarta. Lower: Mitla, in Oaxaca, sacred site of the Zapotec people.

and singing while banging clothes and scrubbing them on the river stones, their babies bobbing up and down on their backs, heads wobbling from side to side.

On forays into the forest near Tepic to catch and watch birds, we occasionally met charcoal burners. They were using the identical method of obtaining charcoal that was used in Arnside. Earth-covered logs were piled into a heap so that the wood burned slowly without combusting. In both places, it was a dying culture. Occasionally, a group of Huicholes would pass us on the trails. The men wore hats decorated with stuffed hummingbirds hanging down from the brim. Women carried babies on their hips or in slings on their backs. We wondered if our friends from the night we spent in the forest were among them, but they showed no recognition.

Not all was fun. A car accident and a bout of hepatitis made life difficult. Years later, I found out through antibody tests that I had contracted hepatitis A—fortunately not B or D, as I feared—after having my chin stitched up in a local hospital following the car accident. Compounding our problems, Peter's supervisor failed to send money. Having eaten hardly anything for a week, we spent our last few dollars on a telegram to the head of our department at the University of British Columbia, who rescued us by wiring the promised money within a few hours.

I joined Peter on two trips to the Tres Marías islands. We did not travel in a dugout canoe this time but in a small, sturdy boat captained by Julian, a wise man who, like many fishermen, had never been to school. Nevertheless, he knew all the stars and currents, and we felt secure, which was a good thing because a hurricane blew up. Julian told us the anchorage was not good, and we should go to a sheltered bay on the next island, María Cleofas, before the worst of the hurricane hit us. Sheltered in the bay, we saw a boat in distress coming in from a weird angle. Julian signaled the way with lights, and the deluxe boat came in badly damaged. The American man captaining it was alone and said he had been fishing for Sailfish. He pleaded with us to tow him back to Puerto Vallarta the next day, promising to pay for everything. None of us, especially Julian, trusted him. Why was he alone, and why was his boat so badly damaged when the hurricane had hardly begun? However, he was clearly in difficulties, so we spent the next day towing him back. We asked only for the cost of fuel. After arriving, he quickly disappeared, without paying for the fuel, never to be heard of again, not even a thank-you. Who was he—a fugitive from justice?

On many islands in the Pacific, fishers, pirates, and buccaneers have introduced goats and sometimes pigs as a living store of future food. María Magdalena had plenty of goats and deer, whereas María Cleofas lacked feral animals. On both islands, birds were abundant and easy to find just by walking up dry arroyos, and I remember my first bird was a Happy Wren duetting with its mate. They were so well synchronized that it was impossible to distinguish which one started and which one finished the song unless I watched the beaks carefully. A few pools of water on María Magdalena attracted not only birds and goats but also snakes. Peter was struck on the boot by a small boa constrictor, and later we watched a massive one and wondered if a baby goat nearby was in danger. This was the first of what would later be many research explorations and discoveries of uninhabited islands together, and what a joy it was.

For the next trip to the islands, we had to visit the penal colony on María Madre to get a permit. This was the one inhabited island of the Tres Marías. The prisoners crowded round the boat, and in fluent English begged us for books in English. "Anything," they said. "We are starved of literature." Were they political prisoners? Their English was impeccable. We gave them the only two books we had, a couple of novels, one by Dostoevsky.

Elsewhere in Mexico we visited archaeological sites whenever possible. Having read as much as we could about early Mexican history, it was exhilarating to see the evidence. On a trip to Oaxaca, we visited Monte Albán and Mitla. This area was the center of the Zapotec culture twenty-five hundred years ago. With its advanced writing and agricultural system, it predates the Mayan, Mixtec, and Aztec civilizations. The Zapotec had a hieroglyphic writing system with a separate glyph for each syllable and two calendar systems, a sacred one of 260 days partitioned into thirteen subdivisions and a civil one of 360 days divided into eighteen months. Maize, beans, squashes, and chili peppers were grown on elaborately irrigated terraces. I was most intrigued by the geometric wall decorations of cut and polished stones in Mitla. They resembled the motifs on Greek vases, symbols of unity and infinity, and indeed they are referred to in Mexico as *Grecas*. One cannot help wondering whether the intended meaning was similar. The pattern is unique, not found in any other Mexican archaeological site, as far as I know. One of two official explanations is that they refer to woven patterns, and we did see the same motifs on serapes and wicker baskets in the market. I prefer the other, not mutually exclusive explanation, which is that, in flickering torchlight at night, the walls appear to move, ushering a person through life to their eternal resting place. Mitla was the Zapotec religious center and place of eternal rest.

We revisited Mexico City sixty years later. Return visits to Teotihuacan, the archaeological site; Tenochtitlan, the capital of the Aztec Empire within Mexico City; and the Museo Nacional de Antropología, with its incredible collection, made us appreciate the extent of the excavations and new knowledge that had been acquired in the intervening years.

Back at UBC, I continued my two jobs and enjoyed participating in many biological discussions with graduate students, postdocs, and faculty. Because I had been trained in genetics in Edinburgh, I was asked to give a talk to this group on the cracking of the genetic code, a hot topic and one with special meaning to me after Peter had whispered the news at the altar. I was able to glean all the necessary information from the excellent university library. In 1951, James Watson, Francis Crick, and Rosalind Franklin had discovered that DNA is a double helix able to both produce exact copies of itself and carry genetic instructions. However, there remained a problem: How do four DNA/RNA nucleotide bases (adenine, or A; thymine, T [uracil, U, in RNA]; cytosine, C; and guanine, G) produce twenty amino acids, the building blocks of proteins? If the code was one base long, only four amino acids could be made, if two, only sixteen, and so on. In 1961, Marshall Nirenberg and J. Heinrich Matthaei had the insight to make a synthetic RNA chain of units of uracil and found that the code for the amino acid phenylalanine was UUU. A three-base code produces sixty-four combinations. They later showed that sixty-one were used for the twenty amino acids and three for stop codons, necessary for the formation of a polypeptide chain. The fact that more than one triplet codes for the same amino acid, they argued, minimizes errors in the formation of a polypeptide in the production of a protein.[2] This was a breakthrough in understanding how enzymes are produced and was foundational for future genetic research. My talk went well, and afterward, several faculty members offered to supervise me for a PhD.

I loved discussing science with Peter and helping him but still hankered to do my own PhD. By now the Icelandic charr project was out of the question. Instead, I could do a PhD at UBC, but that would be difficult because it would cost money, and we relied largely on my income. Moreover, Peter would finish before me. Another question was when to have children. My inclination was to wait until after I had a PhD, whereas Peter thought it better to have children early to minimize the chance of birth defects—a valid reason. We decided on that strategy and that I would do my PhD later, at the first opportunity, when the child or children were in school. I soon became pregnant but, sadly, lost the baby after a six-month pregnancy.

In 1965, after only three years, Peter completed his PhD and obtained a postdoctoral position at Yale University with Evelyn Hutchinson, whom he greatly admired. We decided to drive across Canada, park our VW at Yale, and return to England for a few weeks before Peter's postdoctoral fellowship began in September. This was our first visit home since our wedding. It was wonderful to see my parents. My dog, Leda, treated me like a long-lost friend, and I had time to show Peter my walks and haunts around Arnside. Peter bonded with my parents, with my mother at first and then in later years also strongly with my father. Peter and my mother shared the same sense of humor, and I often found myself the butt of their jokes. Peter said he found the family he never had, warts and all, and entered it wholeheartedly.

On our first day at Yale, I hunted for a job. My diploma from Waddington's course, and a letter sent ahead unbeknown to me from Professor Cowan at UBC, were magic passes—by lunchtime I had five offers of research assistant jobs to choose from. I chose to work with Sheila Counce, who had received her PhD from the Institute of Animal Genetics in Edinburgh and was then working on sex-linked lethal genes in insects. This was her last year at Yale, as she and her husband, Bruce Nicklas, were leaving for teaching and research positions at Duke University. Sheila was a strong supporter of women scientists, skillfully employing quiet, polite diplomacy to achieve her goals. I loved discussing recent genetic and other scientific research with her. Again, I had two jobs. The other was a teaching assistant in two of the first-year labs. At that time, all Yale undergraduates were men, and I often wondered what they thought when a young pregnant female—I was pregnant again—was instructing them. I had the opportunity to go to excellent seminars in the Zoology Department, join discussions on the latest papers, and enter the intellectual life of the university. I particularly enjoyed Evelyn Hutchinson's discussion groups; the topics ranged across a broad field, from questions concerning the possibility of extraterrestrial life to how populations of *Daphnia* (water fleas) and copepods (tiny crustaceans) persisted in the presence of predators in local lakes.

Our combined income was $8,500 a year. I earned $4,500, making us wealthy enough to afford a $2,000 hospital bill when our daughter Nicola was born at the end of the academic year. Sheila helped me to find a good gynecologist, Dr. Billings, after a disastrous visit to another man who proposed a drastic treatment using hormones because I had previously had a late miscarriage. When I questioned this man about the drugs he was about to prescribe, he casually said, "Well, if it is born a boy, we can always remove the penis!" I confided in Sheila and, horrified, she got on the phone immediately and arranged an appointment for me the next day with Dr. Billings. Nicola was born July 19, 1965, in the Grace–New Haven Hospital, a whopping eight pounds five ounces, two weeks overdue. She was put in Peter's arms a few minutes after her birth. He set off skipping down the corridor, shouting, "I have a daughter, I have a daughter," with Dr. Billings and a bevy of nurses running after him shouting, "Mind her head, mind her head." I had worked in Sheila's office up to the last day, and it was strange to be suddenly housebound with a baby to care for. Peter had accepted a job as assistant professor at McGill University in Montreal, and we returned to Canada in September 1965 with two-month-old Nicola.

PART III

Motherhood

12

Rosemary's Monday

> I would give wings to children, but I would leave it to them to learn how to fly by themselves.
>
> —ATTRIBUTED TO GABRIEL GARCÍA MÁRQUEZ

In Montreal, I soon met other young families and shared babysitting with Pauline Checkley, who had a daughter one month older than Nicola. Looking after each other's babies allowed us to go to the dentist, get haircuts, and so forth. Peter immediately settled in as assistant professor. We made good friends with Jaap (a limnologist) and Evelyn Kalff, who arrived at the university at the same time with one baby. We found a small and not too expensive apartment on Ridgewood Avenue within walking distance of the university.

Old, established, and still tethered in its traditions to Europe, McGill was in stark contrast to the University of British Columbia, a young and vibrant institution making its way in a new world. My initial impressions were soon reinforced when only two weeks after we arrived, Peter announced that I had been invited to the Faculty Club for afternoon tea with the other wives of faculty members of the Department of Zoology. The invitation was from four to five p.m. I was asked to wear a hat and white gloves! Horrified, I refused to go, as I was breastfeeding Nicola; moreover, I had neither hat nor gloves, and I was not going out to buy them. Peter pleaded, "It is my first job, and the wife of the head of department invited you so it must be important." My mother had just arrived to see Nicola. She had a hat and gloves, and together she and Peter persuaded me to borrow them for the occasion. Evelyn was in the same dilemma. We decided to go together.

On the way there she said, while driving, "Have you any idea what this will be like?"

"Yes," I said. "We will all sit round low tables, a waiter will bring in cucumber sandwiches cut in triangles with the crusts removed on a silver tray, and the wife of the head of the department will pour the tea."

"How on earth do you know?"

I replied, "This is the way the old ladies in Arnside do it!"

She asked if we were to take off our gloves to eat the sandwiches. There she had me stumped; we had better watch the others, I decided. To our amazement, it turned out to be exactly as I had described, even down to the triangular cucumber sandwiches on a silver tray. Trying desperately hard not to laugh, I caught Evelyn's eye, and she started to shake with laughter, brought out a dainty lace handkerchief, and pretended to sneeze. This was 1965; the Faculty Club was for men only, and that day was the one day of the year when women were allowed to enter. The club had no women's restroom, ensuring the tea party lasted no more than one hour. I got home to find poor Nicola desperate to be fed and my mother trying to comfort her. In the meantime, Jaap and Peter had gone to the pub for a drink—or two or three!

When Nicola was eleven months old, we had another chance to visit England. I was delighted that the cheapest way to get there was to fly Loftleiðir, now known as Icelandic Airlines, as it gave us the opportunity to visit Iceland and to get a glimpse at the habitat of the charr. The flight attendants swore that our fair-haired daughter was Icelandic and played with her at the back of the plane the whole night after wrapping us up in Icelandic blankets and giving us a stiff brandy. I got worried only when they teased us by taking her through a different line at immigration and customs. We took a bus to Mývatn (Lake of Midges); a lake with boiling mud, fumaroles, many Harlequin Ducks, and a colony of terns that swooped alarmingly low over our heads. Fortunately, the midges were not in evidence, or at least they were not biting in the cool temperatures. We passed fjords that I imagined were full of charr, a momentary snapshot of what it would be like to do a PhD there. Nicola was as impressed as we were with the scenery. She was an early talker. To her, sheep were "peesh."

Thalia, our second daughter, was born the following year. We returned to Iceland when she was two and Nicola was almost four. This time we stayed in a small village, Vík í Mýrdal, situated on a broad and black volcanic beach close to the cliffs where puffins nested. Einar Arnason, Peter's graduate student, was interested in the genetics and patterning of *Cepaea* snails in this area.[1] Einar later went on to study genetics at Harvard and then became a professor at the University of Reykjavík. Peter was interested in the effects of puffin feeding behavior on survival of their chicks. We watched them being harassed by skuas (*Stercorarius*) as they flew in from the sea, each puffin with about six sand eels dangling from its beak. Dropping one distracted the skuas and resulted in a frenzy as they dived in a tangle to catch it before it was lost to the sea. The puffin in the meantime flew on to its burrow in peace with food for its single nestling.[2] We were to see this behavior, termed *kleptoparasitism*, many years later in the Galápagos, where frigatebirds chase and harass tropicbirds and boobies as they bring fish back to their nestlings.

In 1969, Vík was a tiny village with only a few houses. Twelve young children of varying ages would knock on our window and ask if Thalia and Nicola could come out to play on the beach. One rainy day I found them sitting in a shelter, open on one side, with a bench against the other three walls. The children were telling a story, each one contributing a piece before passing it on to the next child. I thought of the oral literature that led to the sagas. I wondered what would happen when it was Nicola's and Thalia's turn. Nicola, not knowing one word of Icelandic, had acquired the rhythm of the language and intoned *lar-de-dar, lar-de-dar-de-dar* for a minute or two in perfect tempo and cadence, complete with arm gestures. Thalia, picking up the cue from her elder sister, did the same, and so the tale was passed on to the next Icelandic child, without a flicker of surprise from any of them. Perhaps a new language is learned in this way, through a combination of copying and extemporizing. It was intriguing to later learn from Erich Jarvis's research that rhythm and arm and body gestures are closely associated with the speech centers in the brain.[3]

Some of the older children, ages ten to twelve, said they had learned English by watching television and reading the subtitles. Television and radio programs were strictly controlled by the government. Taxi drivers circumvented these restrictions with long aerials attached to their cars to pick up the radio relayed from the American base at Keflavík. The older children impressed me by knowing the Latin names of all the minute plants at the back of the beach and were a fount of knowledge about the geology of the area. There was no school, but as had happened for generations, the children went from family to family, learning different subjects while helping with the farmwork.

Fifty miles offshore from Vík was Surtsey, an island formed from an underwater volcanic eruption in 1964. The eruption continued until 1967, and by the time we saw it from the air in 1969 as we landed in Reykjavík, it was 0.59 square mile in area and 588.5 feet above sea level. The islet was immediately made a nature reserve, accessible to only a few scientists allowed to visit to document the colonization of plants and animals. Mosses, lichens, and flying insects were the first to establish residence. In 1998, the first bush, a willow (*Salix phylicifolia*), was found, to be followed by two to five new plants each year. By 2009, gulls, puffins, and a golden plover were nesting there, but at the same time, the island was sinking due to subsidence and erosion at the coastline.[4]

Thalia's year of birth, 1967, was also the year Canada turned one hundred. People flooded to Montreal to see Expo 67, the international exhibition held to celebrate the centennial. The theme, Man and His World, reflected Canada's interest in global diversity. It was divided into Man the Creator, Man the Explorer, Man the Producer, Man the Provider, and Man in Society ("Man" being short for humankind, I hope!). Among the international pavilions, the most outstanding were Buckminster Fuller's geodesic dome from the United States, a display from Czechoslovakia illustrating the making of glass, and Moshe Safdie's Habitat, a series of modules in single buildings offset in such a way as to provide privacy, air, and light, and with sufficient space for growing plants, providing the inhabitants with a cheap and pleasant way of living. There were performances by La Scala; England's National Theatre, with Laurence Oliver in *Uncle Vanya*; the Amsterdam's Royal Concertgebouw; and many others. We were inundated with visitors that year, all wanting a bed or floor space.

Our two-bedroom, one-bathroom apartment was full of families for six months. At one point we escaped to the peace of the university's field station at Mont Saint-Hilaire and left our guests to their own devices. There were several lasting effects of Expo in Montreal. One noticeable and amusing one was that the bus drivers, who had been notoriously gruff and rude, had been given a crash course in being polite and helpful. Overnight, it seemed, they became friendly, and most striking, they seemed to be enjoying their change of mood.

I loved being a mother but missed the intellectual stimulation of academia. I asked if I could use the McGill University Library and was warmly welcomed by the librarian and staff. Peter and I decided to hire a babysitter one morning a week so that I could use that time in the library to catch up on scientific literature. It became known as Rosemary's Monday. I continued with these Mondays throughout the children's preschool years and refused to use the time for anything other than keeping up with research, so I would not be too far behind when the chance came to continue my career in biology, which I hoped would be possible when the children were in school. Many years later, Peter and I were giving talks in Arizona, and while there I was asked, as frequently happens, to give a second, impromptu lunchtime talk to graduate students about combining family life with science. The next week we had an e-mail message from the department, thanking us for our talks and adding that we might be amused to know that the previous Monday, all the women had taken the day off. When asked why, they said they were having a Rosemary's Monday in the library!

Soon after we arrived in Montreal, Peter began research on competition between field voles (*Microtus*) and bank voles (*Clethrionomys*). *Microtus pennsylvanicus* and *Clethrionomys gapperi* are found in North America, while similar species in the same genera, *M. agrestis* and *C. glareolus,* occur in the United Kingdom. He had collected live animals in the woods around Arnside and on the Welsh island of Skomer, keeping them in my parents' attic before exporting them back to his laboratory at McGill. When Thalia was a year old, we drove as a family to Newfoundland to collect mice for Peter's experiments. While there, we made a long diversion to L'Anse aux Meadows at the far northern tip to visit the newly excavated Viking site from the eleventh century. We were the only visitors and were fortunate to meet Dr. Helge Ingstad, who with his wife, Anne Stine Ingstad, had discovered the site in 1960.[5] Only Helge was there at the time. He gave us a personal tour, enthusiastically explaining how the remains of the buildings perfectly matched those found in Greenland and Iceland in the tenth century. Discoveries of a gilded linchpin, part of a brooch, remains of a loom, a spindle, and knitting needles all suggested that the site had been a settlement that included women. In the Icelandic *Saga of Erik the Red,* this settlement is mentioned as Vinland after the finding of grapevines by Leif Eriksson. Wild grapevines (*Vitis riparia*) and butternut trees (*Juglans cinerea*) grow in abundance on the banks of the Saint Lawrence estuary, to the south. Butternut husks were found in middens in Vinland, leading Helge to reason that the inhabitants had made forays down the coast to the mouth of the Saint Lawrence River, where they obtained both grapes for making wine and butternuts. I have always been fascinated with the movement of humans across the globe and how groups with different backgrounds mingle, exchanging genes, language, and culture. Thinking along these lines, I asked Helge if they had found any evidence of contact with the local Indigenous residents, but he had found none.

We know now from the rings in the fir and juniper wood used to make the buildings that the wood was from trees felled in 1021 CE. This astoundingly accurate date was made possible by finding a carbon-14 spike known to have occurred in 993 CE from a burst of high-cosmic-energy particles (referred to as a Miyake event) and counting the rings from that point.[6]

When we returned to Montreal, a letter was waiting for us. It was a questionnaire from the British government sent to all students who had graduated from a British university in 1960. Many questions centered around what we were doing now, in 1969. Peter and I dutifully answered them and sent them back. This turned out to be ammunition for the 1971 Rothschild Report, which, among other things, undercut education for women. Although I had worked up to 1965, I was now a mother of two young children. Without daycare help, I was looking after them at home, and in British terminology I was a "housewife," and I ticked that box. There was no box that said this would be a short interruption to my long-term professional plans. A summary I received from the government survey gave the percentages of women who were now mothers and housewives, and the main British papers referred to the economic waste of educating women at university if all they did was return to these domestic roles. The hurdles for women had, at one stroke, become mountainous. Peter and I were furious at having inadvertently helped to create them and vowed to never again respond to a government survey.

The Rothschild Report claimed that funds from research councils should be controlled by the government and be commissioned on a "customer/contractor" basis. In Victor Rothschild's words, "The customer says what he wants; the contractor does it (if he can); and the customer pays." He wrote numerous articles in the major daily papers referring to himself as a rationalist and his ideas as "common sense and self-evident." The government loved it, but there was an uproar in the Royal

Society, led by Peter Medawar, who pointed out that Rothschild spent most of his time in commerce and did not understand or appreciate the value of creativity in pure scientific research.[7] Rothschild should have known better because, as an embryologist, he spoke the same language in science as Medawar, an immunologist. Rothschild, in collaboration with Michael Swann, the University of Edinburgh professor who had encouraged me to go to UBC, had demonstrated that the first sperm to fertilize an egg causes a change in the proteins on the egg's surface, thereby preventing penetration of additional sperm. Rothschild applied this discovery to the cattle industry. In addition to his scientific endeavors, he was a banker, a senior executive of Shell oil and gas company, and an adviser to the politically conservative governments under Prime Ministers Edward Heath and Margaret Thatcher. It was a foregone conclusion that his report would be accepted by the government. Perhaps someone should have sent him Marie Curie's statement:

> *We must not forget that when radium was discovered no one knew that it would prove useful in hospitals. The work was one of pure science. And this is a proof that scientific work must not be considered from the point of view of the direct usefulness of it. It must be done for itself, for the beauty of science, and then there is always the chance that a scientific discovery may become like the radium a benefit for humanity.*[8]

Peter and I threw ourselves into being encouraging parents. Peter firmly believed that children should have a complex and enjoyable environment and be introduced to a diversity of experiences. In Montreal that was easy to accomplish, as there was an excellent children's library in a nearby park and not just English- and French-speaking children to play with but also those of families from throughout the world that had emigrated to Canada. What our urban life lacked was access to wilderness.

McGill's field station at Mont Saint-Hilaire, about twenty miles east of Montreal, satisfied this and became our second home. It gave the children opportunity and freedom to explore the local wildlife. Visible from miles away, Mont Saint-Hilaire stands alone on a flat plain and is one of the Monteregian Hills, formed by plutonic intrusions as the North American Plate passed over a hot spot during the Cretaceous period. A magnificent forest of maples and beeches, with some trees over four hundred years old, covers the slopes surrounding a large central lake (Lac Hertel) in a glacially formed depression. The land had been bequeathed to McGill University by Andrew Hamilton Gault in 1913, whose large dwelling, known as Gault House, was still situated on the property. In the 1960s, this large natural area was hardly used. Peter jumped at the opportunity to hold an active field course in ecology and develop many student projects there. While he was teaching, the children and I lived in a large cabin that had been left empty. The university was happy to see it occupied for long stretches at a time. Nicola and Thalia swam and canoed in the summer and skated and skied in the winter. Each winter, ice fishers would arrive at the lake, drill a hole in the ice, sit in a small tent, and patiently fish for hours. Their reward was quite a few pike to take home for dinner.

Herman Smith-Johannsen, better known as Jackrabbit, a legendary old man who had introduced cross-country skiing to Canada, frequently stayed in Gault House with his daughter, who was custodian of the field station. He was then in his nineties. We would often encounter him, and he never failed to give the children skiing tips. One day he said to them, "If you get stranded in the mountains, do you know how to survive? Do you know how to start a fire? Follow me." We all trooped into

FIGURE 7. Upper: Peter with Nicola and Thalia. Lower left: Nicola meeting her newborn sister. Lower right: With Nicola at Mont Saint-Hilaire, Quebec, 1966.

Gault House, the children following him wide-eyed. Bending down, the hoary old man kept the two little girls entranced as he placed a small pile of dried moss and hay on the stone slab in front of the fireplace, took out a flint, scraped it, and created sparks. Gradually, the moss caught fire. "Never" he said, "go out for the day without a flint and dry moss or hay." The two girls nodded solemnly. He never tired of telling Nicola that he was exactly ninety years older than her. He was an amazing man, always interested in other people and things around him, often peppering the children with questions. Having obtained an engineering degree from a German university, he retired at age fifty and lived to 111. He died on a visit to his original hometown in Norway. Passionate about introducing the wilderness to young people, Smith-Johannsen published a free bilingual book of topographical maps showing ski trails and hiking trails in the Laurentian Hills north of Montreal, many of which he had blazed himself. He established a two-day, hundred-mile ski marathon from Lachute in Quebec to Ottawa, and at age one hundred skied the first ten miles, after having broken his leg at ninety-nine years old. Remarkable!

We drove twice cross North America, camping along the way, introducing the children to the geology and the wildlife. Peter would write in the morning, and together we would go for long hikes in the afternoon. The children delighted in chasing butterflies in the meadows and watching hummingbirds. On one occasion, Thalia (age three) accidentally caught a hummingbird as it swooped down in its display flight straight into her outstretched open net. I don't know who was more surprised, Thalia or the hummingbird. The girls painted with natural orange, red, and yellow ocher clay that they found in a hollow in some muddy rocks and hunted for fossils and lumps of agate. Sightings of pikas, Mountain Goats, and Bighorn Sheep were

FIGURE 8. Upper left: Peter with Nicola on his shoulders. Upper right: Thalia on my shoulders. Lower left: Thalia and Nicola on Peter's back. Lower right: Skiing at Mont Saint-Hilaire.

some of the highlights of these exploratory trips, as well as toasting marshmallows around campfires. In the United States, we visited the Grand Canyon, Death Valley, Meteor Crater, and the Chiricahua Mountains. We were enormously impressed by the museums in each of the national parks and the valuable information they provided on geological formations, animals, plants, and other life-forms of the local areas, clearly explained for young children as well as adults. Thalia was in her earliest fossil-finding phase, and she seemed to have an uncanny ability to find tiny fossil trilobites in streams, very much helped by Nicola.

13

Thwarted

> Life is like riding a bicycle. To keep your balance, you must keep moving.
>
> —ALBERT EINSTEIN

By 1971, Peter had received tenure and been promoted to full professor at McGill and was due for a sabbatical leave. He had just started a new research project, further exploring a specific example of character displacement (divergence) and its evolutionary causes. The problem focused on *Sitta neumayer* and *Sitta tephronota,* two closely related species of nuthatches. *Sitta neumayer* occurs in Greece and Croatia, *S. tephronota* in Afghanistan, and the two species overlap in Iran, where they differ from each other in body size, beak size, plumage, and possibly diet and song, much more than they do in isolation. Peter wanted to test the idea that interactions between the species had caused the divergence in these traits, an idea that originated with Darwin.[1] He chose to spend the year in Oxford and to travel from there to Iran, Turkey, Afghanistan, what was then Yugoslavia, and Greece.

I looked forward to this year. Both children would be at school for the first time, giving me the opportunity to initiate a research project of my own that would lead to a PhD. I was very keen to visit Persepolis in Iran, having read extensively about the Achaemenid Empire, which had existed between 550 and 330 BCE, and to see firsthand part of the Fertile Crescent, where domestication of animals and plants and writing had originated. Trade routes into the area had led to exchanges of ideas between peoples and to innovations in technology and mathematics. Eventually, degradation of the land through overuse depleted the soil, forcing people and their livestock to expand outward. I reasoned that this diaspora must have produced interesting phenotypic changes—that is, observable changes due to the interactions between genotype and environment in animals and plants that were domesticated, as well as to the native flora and fauna that would have been impacted by agriculture. I considered this question for my own research. One example was in the many changes that had occurred during the domestication of sheep. There were still wild Mouflon, the species that gave rise to the domesticated sheep, in game reserves near Shiraz.

Dr. Squires, our doctor in Wantage, a town near our rented home south of Oxford, gave us all inoculations in preparation for the trip and told us he had spent several years in Shiraz, the city that was to be our home base, with his young family. Our initial plan was to drive to Tehran through Turkey, but we were warned, first by our neighbors in Montreal who had worked in Istanbul, then by the Canadian Automobile Association and our car insurance company, that traveling by car through Turkey would be prohibitively expensive and could be dangerous. Peter decided to fly to Iran on his own, for a month. He was still hoping that we might all be able to go there later as a family, if we could acquire extra car insurance in Istanbul.

I was less optimistic about our change of plans and feared my decision to relaunch my research career when both children were in school was on hold. The year was stimulating for us all as a family but frustrating for me in terms of my progression toward my own research. Even the library was unavailable, being reserved, understandably, for Oxford professors and students. The month Peter was in Iran was a month of deep contemplation. How was I to combine raising a family with pursuing an intellectually stimulating career? The thought-provoking visit to Iran was in jeopardy. Should I switch careers from my intended PhD in biology and perhaps link biology with archaeology, geology, or medicine? The charr problem in Iceland did not seem compatible with a family, particularly if Peter's research was elsewhere. However, there were many similar interesting questions and problems to consider that might be more amenable to balancing a family with research.

The week before Peter came back from Iran, I returned from collecting the children at school to find a policeman waiting for me. With a slight smile, he said, "You are under arrest, you are driving illegally." This was the last thing I needed! I was about to show him my driver's license, when he said, "No, your car's license plates are out of date." Apparently, the Volvo, which we had purchased in Sweden and would take back to Canada, could remain in England for only six months, and that time was up. He left me, saying, with an enigmatic smile, "Think of a solution; here is my telephone number." After wondering what I could do, I phoned him and said, "What if I take the car over to France for a day and reimport it?" He laughed and said, "Yes, but I did not suggest it."

I promptly arranged an interesting trip to France for the weekend after Peter returned from Iran. We had visited the megalithic stone circles of Stonehenge, Avebury, and the oldest one in the United Kingdom, dated to 3000 BCE, Castlerigg in Cumbria. I had read Alexander Thom's fascinating if somewhat controversial account of other megalithic monuments across Europe and his ideas that they acted as lunar and solar observatories (chapter 6). The Carnac stones in Brittany were within a day's drive and had a campground nearby. Reimporting our car from France would be easy and legal and give us the opportunity to visit this interesting area. Three days after Peter returned, we were on a ferry to France.

The seven-thousand-year-old stones at Carnac were aligned, we read, in the direction of the sunset at the solstices and were surrounded by pre-Celtic burial barrows. The placement of other giant stones was calibrated with the lunar cycles. Thom theorized that the Carnac and other giant stones were astronomical observatories arranged and spaced according to the megalithic yard (2.72 feet).

It was cold and raining heavily. After a tour through the burial barrows holding lighted candles dripping hot wax, we were invited to have dinner by a roaring fire in the kitchen of the campground farmhouse before crawling into our cold, damp tent. While waiting for flaming *crêpes bretonnes,* Nicola and Thalia played with three large dogs lying in front of the fire. Nicola, who refused to speak French in England, switched into perfect French when playing with them. Her argument was, "Well, of course, we are in France, the dogs are French!"

A month later, we drove to Greece and on to Istanbul, in the hope of acquiring the extra car insurance to allow us to drive through Asiatic Turkey to Iran. While driving in Turkey toward Istanbul, we saw two dangerous accidents in quick succession. On our right was a campground, and we immediately drove in, deciding to take the bus into Istanbul the next day. The campground was heavily guarded by military police. In the middle of the night, the children wanted to pee, and I was just about to take them to the bathroom, when suddenly two rifle barrels pointed at us through the tent door. The police were very friendly once they realized our situation, and to the children's amusement, we had an armed escort to the toilet and back, with my heart still pounding.

An ancient wall surrounds Istanbul, and at the city gates there was a man leading a bear walking on its hind legs, a chain around its neck, horribly reminiscent of medieval bearbaiting. We walked along the Galata Bridge, which spans the Golden Horn, seeing men smoking hookahs. They reminded the children of a picture from *Alice in Wonderland* in which the caterpillar sits on a mushroom smoking a hookah before his metamorphosis. "Is the smoke dangerous?" they asked. Next, the spice market, with exotic aromas emanating from sacks of different-colored spices, then on to Hagia Sophia. I had read about the amazing dome, built on arches placed at each corner of a square. Layers of arches spiraled upward to a final ring of arches through which the sunlight was streaming, rays bouncing off gold plate, creating the illusion of a floating dome. I stood mesmerized, thinking how this feat was the creative recombination of all previous attempts at domes by past architects, a possible example of cooperation between peoples from different cultures. "Mummy, come and look at this." The children brought me down to earth to watch art restorers gently scraping a wall to reveal the original mosaics from 537 CE.

In between excursions to these sights, we visited the car insurance company and had it confirmed that driving east across Turkey was out of the question. It was not unexpected and the reason I had initially hoped we would all fly to Iran the month before. Our chance in a lifetime to visit the Khyber Pass, and mine to Iran, had gone, leaving me bitterly disappointed but with the determination to think of a way out of our impasse.

Instead, we drove to northern Greece, where one of the nuthatch species, *Sitta neumayer*, occurs in the absence of the other, *S. tephronota*. We found a perfect campground in Kavála in northern Greece, close to the nuthatch's territories and on the Mediterranean Sea. On our first walk up a steep rocky slope through aromatic Mediterranean herbs, a small tortoise crossed in front of us. It was the children's first wild tortoise. Then, above us, on the rocks, I heard my first living nuthatch, an evocative series of downward notes that seemed to echo off the surrounding cliffs, reminiscent of the song of the North American Canyon Wren, which lives in similar habitat. The nuthatch's nest, decorated with a snakeskin and bits of colored paper, was placed under a rocky overhang nearby.

During Peter's daily excursions he discovered many nuthatch territories, which enabled him to record songs, perform playback experiments, and document feeding and other behavior. The children loved being there, gained a few Greek words, and learned to swim. They attracted the attention of local fishermen, who were thrilled that they had Greek names and would call out "Nicola, Thalia," and tell them to hold out their skirts to form a bag that they then filled with small octopuses and fish for dinner. One day, a strong thunderstorm sent us running into the tent, where I read the girls Greek myths and legends to distract them. When Peter arrived back, drenched, some hours later, Nicola greeted him with, "Daddy, Zeus must have been very angry with you." Many years later, Nicola read her own two children the Greek myths from the same book.

It was the time of the far-right Regime of the Colonels. There was no evidence of this military regime in Kavála, where we were impressed by the irrepressible exuberance and passion for life that Nikos Kazantzakis had described in his book *Zorba the Greek*. Perhaps we were lucky. The camp manager told Peter that his office was visited each night by a military man, who tried to listen to the tape recorder that Peter had left recharging on his desk. He never did realize that you had to press two buttons simultaneously to turn it on. We were concerned because the recorder was picking up interference from a local radio station while recording birds and, for all we knew, possibly police radio communications.

On our way to Athens to visit Peter's half sister Sarah and her new baby, Nicholas, we camped in Delphi on Mount Parnassus. Dedicated to Apollo, Delphi is perhaps most famous for the oracle where Pythia divined the future outcome of major events. Thalia and Nicola were more interested in the display of butterflies flitting about the ruins and their first sight of glowworms at night. Continuing south, we visited the ancient theater in Epidaurus, which we had to ourselves. The children tested the amazing acoustics, with one sitting in the upper back row and the other dropping a coin on the stage. Apparently, the limestone seats filter out low frequencies, allowing voices, and in this case the sound of a dropped coin, to carry. After visiting the impressive Acropolis, Parthenon, and museums in Athens, we then traveled to Sarah's house in the Peloponnese, the ancient homeland of the Spartans. I told the children that in 400 BCE it was better to be a woman in Sparta than in Athens. Spartan women were allowed to live a life like a man, own property, and have equal education, all barred to Athenian women at that time.

We squeezed in a short visit to Crete. Instead of driving our car onto the ferry, we stood on the quay and watched as it was winched up from dock to deck. We were told our insurance only covered it while on the deck or on land, so we stood anxiously watching as our car dangled, lopsided and hesitatingly, over the sea for what seemed ages, not knowing whether to pray to Poseidon or Zeus! Our destination was the Minoan palaces of Knossos and Phaistos, evidence of the earliest, most advanced civilization in Europe. We visited both and the archaeological museum in Heraklion. We were not disappointed. As we discovered from the ruins and museum, Knossos must have been a magnificent multistory building surrounding a large courtyard with storage room for grain. Separate pipes for sewage and clean water showed a heightened level of sophistication, confirmed by paintings indicating a knowledge of agriculture, with certain crops cultivated together. The exquisitely realistic, dynamic frescoes of people leaping over bulls, flying dolphins, flowers, olives, fish, and octopus were extraordinary. Some looked very Egyptian, particularly a line of young men depicted sideways and bearing gifts and women pouring wine. In contrast, the figures of women with wasp waists and elaborate dresses that revealed their breasts seemed uniquely Minoan, as did a natural portrayal of a woman collecting saffron. Frescoes of ships laden with goods suggested a flourishing trade and could have explained the possible Egyptian connection. The Minoans' written language, Linear A, was impressed on many clay tablets, and on the Phaistos Disc. It has never been deciphered, unlike Linear B, which was derived from Linear A by the Mycenaean Greeks around 1400 BCE.

The Minoans seem to have been a peaceful, trading, open society, as there is no sign of wars or armies. A massive earthquake and tsunami in 1560 BCE that occurred on Thera (now Santorini) is recorded in tree-ring data not only in Europe but also across the world. This event apparently destroyed Knossos and other Minoan palaces, but the Minoans recovered and rebuilt them. In 1450 BCE, the palaces were destroyed again, possibly by another earthquake. Shortly afterward, warring Mycenaean Greeks invaded from the mainland, possibly taking advantage of the chaos in the aftermath of the natural disaster.

We returned to Europe through Yugoslavia. There we entered a different world. Tito was in power and ruled over this communist country, keeping Serbs, Bosnians, Slovenes, Croats, and

Montenegrins in temporary, uneasy peaceful coexistence. Judging by the numerous hotels and campgrounds, now empty in May, the tourist industry was a major source of income later in the summer. We drove on narrow roads and passed farms with wheat fields full of red poppies on either side. Today, I cannot help thinking of John McCrae's 1915 poem "In Flanders Fields," a forewarning of the bloody devastation that occurred in the region only a few years later, in the 1990s. Simmering ethnic conflicts exploded into a full-blown war in Bosnia between 1992 and 1995. In 1971, however, all was peaceful on the surface, with evidence of an apparently bountiful harvest. Seeing all this lush produce in the fields it was a surprise to find rather tired-looking vegetables and fruits in the markets, until we were told that all food was sent to Belgrade or Zagreb and then redistributed.

Our destination was a campground outside Kotor, a coastal town in Montenegro backed by limestone mountains, which provided perfect habitat for nuthatches but also a perfect firing range for military practice. On the first day, bullets ricocheted off the rocks around Peter, forcing him to redefine his boundaries for nuthatch observations. While we were at Kotor, the pound sterling was devalued, and it became impossible to cash our traveler's checks. Fortunately, we had cashed enough when we entered the country and had stocked up on food. Our camp neighbor, George, who also relied on traveler's checks, was penniless. We fed him for a week, and in return he regaled us with stories of his extraordinary life. An orphan in England, he had been brought up with ten other orphans in a circus. He was the lone survivor, the other nine having died of disease, malnutrition, or falling from the high-wire without a safety harness. A confirmed Communist, he was allowed and even welcomed into Albania and was on his way back from a holiday there to resume his job in customs and immigration in Southampton.

We made our way home through northern Italy. Our last campground in Yugoslavia was in Split, where we hit the beginning of the summer tourist invasion. Leaving early, we entered the Lombardy plain and started our long drive toward Switzerland. We had hoped to visit Cremona on the way, enticed by museums showing the famous Amati and Stradivari violins and violas, but we had no money other than the nonfunctional traveler's checks. We managed to change the last of our Yugoslavian dinara into Italian lire, and our reward was a wad of notes that visibly shrank every time we filled the car with gas. Reaching a campground at dusk, we turned out the car to find enough money to pay for a campsite. From under the back seat, we unearthed a mixture of currency—Austrian shillings, French francs, Italian lire, and UK pounds—that the children had been playing with. The owner, a compassionate, comfortable woman accepted it all and a half hour later appeared with a plate of delicious homemade ravioli that she kindheartedly said were "leftovers" from her dinner.

We left early the next morning for Switzerland, where we knew we could cash our traveler's checks. The banks opened painfully late. At last, we got our money and then a very welcome late breakfast. We made a slight detour to the alpine meadows near Zinal for Nicola's seventh birthday. The visit was timed well, as flowers were at their peak and the sun was shining, while Weisshorn mountain towered above us on our long hike. A wild baby Chamois followed us and gave us up as companions only when we returned to the village.

After a long time away, we went back to Arnside and then returned to Montreal.

PART IV

Return to Research

14

One Step Sideways

We cannot direct the wind, but we can adjust the sails.

—ATTRIBUTED TO BERTHA CALLOWAY

How could I begin my PhD research? Back in Montreal, Peter and I put our heads together. We were short of money and to do a PhD was costly. However, there was a possible solution—for me to do a one-year teachers' training course, get a job teaching school, and start my own research during the school holidays. The cost of this course, at McGill, was made possible by a large deduction owing to Peter's position on the biology faculty. This move turned out better than I had expected. Thanks to my previous experience teaching at Yale and the University of British Columbia, I obtained a position at a local private all-girls school during the final months of the teachers' training course. I was used to holding two jobs, and now I had three: the training course, teaching, and our children, who in terms of priority were number one. Miss Edgars and Miss Cramps School (ECS) was two minutes' walk from Nicola's and Thalia's school, making it convenient to take them to school and collect them at the end of their day. Time became valuable. I amazed myself by becoming highly efficient and organized. I did all assignments ahead of time because I knew I would have to drop everything if Nicola or Thalia became ill.

Nicola had been to a French-language school at age four, and now both she and Thalia went to a bilingual school, The Study. The school motto proudly proclaimed *Le monde a besoin de femmes formidables* (The world needs great women). It became more fully bilingual after they left. Nicola was already fluent in French. Thalia gained enough at The Study and later with an excellent teacher in Ann Arbor, Michigan, to translate and put into context a wide body of information from French manuscripts on the natural and general history of the Galápagos.

Both girls were fun-loving, bouncy, and bright, loved reading and exploring nature, and enjoyed camping. We were lucky. Our small apartment was frequently filled with their friends, who often stayed for a meal and sometimes joined us on weekends. Deborah Garson (known as Borra) was Nicola's special school friend. She was the daughter of the children's violin teacher and an exceptionally talented violinist herself, having performed on BBC Television at the age of three. We became good friends with her parents, Alfred and Crystal Garson. Alfred was a music professor at McGill, and Crystal was about to start her PhD research in psychology.

When Nicola and Borra were seven years old, they had to do a school project in pairs on any animal they chose. Most children opted for horses, cats, or dogs. Borra and Nicola, being imaginative, lively, and cheeky, chose sea squirts. Many telephone conversations took place between the two of them about sea squirts. Since our phone was in the kitchen, I could hear one side of the conversations while I cooked dinner. In one of them I heard Nicola say, "Borra, we have to do something about their fertilization." Silence. Then from Nicola, "No, Borra, not the stuff farmers put on fields." After this, Borra's mother felt strongly that children should be given sex education early, and she casually left a book on the coffee table in their living room that began with the birds and the bees. Of course, Borra and Nicola read it, and this led to some amusing outcomes. Borra's parents were holding a rather fancy cocktail party, and she was asked to hand around the nuts. One prim old lady said to her, "What a nice little brother you have," to which Borra replied, "Yes, I guess my dad must have given my mum a really good seed." Once Nicola, in the middle of dinner, just as I was bringing in a hot dish from the kitchen, casually asked, "When I was conceived, was Daddy on top of you or underneath?" I nearly dropped the dish, rigidly speechless, but Peter just as nonchalantly said, "On top," to which Nicola disinterestedly responded, "Oh," and immediately asked for someone to pass the butter.

Montreal experienced a crisis in 1970, when Pierre Laporte, the deputy premier of Quebec, and James Cross, a British diplomat, were kidnapped by the FLQ (Front de libération du Québec), a terrorist group demanding independence for the province. Pierre Trudeau, the prime minister, invoked the War Measures Act, soldiers were posted everywhere, and we were all under curfew. Pierre Laporte was killed, whereas James Cross was liberated, apparently thanks to the many conversations he had with his kidnappers about their agenda and demands. This is now known as Lima syndrome, a phenomenon in which kidnappers develop empathy for their hostages. The name comes from the takeover of a Japanese ambassador's home in Lima, Peru, in 1996 by terrorists (loosely dramatized in the novel *Bel Canto* by Ann Patchett).[1]

I had been apprehensive about the extremely cold Montreal winters, but they turned out to be enjoyable once we had all acquired the appropriate warm clothes. We skied in the parks and ice-skated on tennis courts flooded for that purpose, and the children built elaborate snow houses and tunnels in the small plot of land behind our apartment, which was on the ground floor of a town house in lower Westmount. This bit of land was visited by many stray cats, all living off rats and bread put out by our neighbor for the birds. The children had names for them all: Socks, Jowls, and Pugilistic Pete, among other colorful ones. A long-haired, fluffy ginger cat they named Gingy stood out from this motley crew. As the winter became colder and colder, Nicola and Thalia made a shelter lined with blankets for the cats, but they wanted the ginger one as their own, and they fed it in the doorway. Peter and I gently but firmly explained that we could not have pets because we traveled too much. Then one day, when we were all at school and Peter was working at home, he turned around to find that Gingy had squeezed through the slightly open window and was sound asleep on his pillow in the bedroom. On returning from school, the children let out whoops of joy on discovering that "Daddy has adopted Gingy!" Well, after a flea treatment and a visit to the vet, Gingy became our beloved pet for the next eighteen years.

Our next-door neighbor said Gingy frequently got into her basement through a half-open window, turned on the television by pushing a knob, and then sat and watched it. I don't know about that—we did not have a television set—but several years later, when we were living in Ann Arbor, we were looking after a friend's dog and a remarkable incident occurred. Gingy and the dog did not get along, and for the first night I put Gingy downstairs and the dog upstairs with a firmly closed door between. Snarls and barks woke me up. Gingy had escaped and was furiously demanding his territory. I separated them again, but the whole performance was repeated, and yet again. How was

FIGURE 9. Gingy.

he getting out? To find out, I stayed downstairs with Gingy. A few minutes later, he stood up on his hind legs and, with his two front paws, turned the doorknob and took a couple of steps backward to open the door; then free, he bounded up the stairs and back into the battle against the poor unwelcome dog. I had read that Somali cats, or whatever mixture he was, were intelligent, and I agree; this is only one of the many occasions Gingy demonstrated he was no ordinary cat.

One outstanding event occurred during my training as a teacher that made a lasting impression on me. We had an extended visit from a woman who had been involved in remodeling the Finnish school system. She told us that before World War II, when Finland was largely dependent on agriculture and forestry, children from farming and working-class families had little access to higher education relative to those from urban and wealthier families. After the war, there was a strong desire for reform to produce social and educational equality at the highest possible level. The reformers' goal was for every child to have access to nutrition, health care, and education, including early intervention for children with learning disabilities. The government first eliminated the track system, which put children on different learning tracks based on test scores, then eliminated all tests except for the final school matriculation examination. The most impressive innovation was a five-year postgraduate training program for teachers, with salary. Entrance into a teachers' training course was highly competitive, and the status, prestige, and salary of a teacher thus became equivalent to that of a medical doctor or a lawyer. "After all" the visiting educator said, "a teacher trains minds while a doctor cures the body." Both are important.

The curriculum was designed for individual children, with emphasis, she explained, on "problem solving, multiculturality, and prevention of learning difficulties." The stated goal was to inspire children to find their individual strengths and encourage creativity. Schools were small, with around three hundred pupils, and the classes were small. Teachers were responsible for developing their own curricula, although there were constant opportunities for them to meet and discuss innovative approaches with each other and with parents. Play was considered important, and students had fifteen minutes of outdoor playtime between lessons regardless of the weather. In the classroom there were no rankings, no comparisons or competition between students. The weakest students, or students that were behind, were given extra help.

Did it work? Finland spends 30 percent less per person than the United States on schooling. Ninety-three percent of Finnish students graduate from high school, compared to 75 percent of

US students. Sixty-six percent of Finns go on to higher education, compared to 38 percent of US students; and Finnish students, unlike those in the United States, do so without accumulating debt, because all higher education for both Finns and international students is free. On the PISA (Program for International Assessment) test for fifteen-year-olds, Finland consistently ranks very highly, occasionally on top. Most Finnish schools have an increasing number of immigrants. Some are from neighboring Sweden, others from Afghanistan, Bosnia, India, Iran, Iraq, Serbia, Somalia, Thailand, Turkey, Vietnam, and today Ukraine. Placing the emphasis on the individual student, a teacher often finds immigrants learn complex subjects such as science best in their own language; then they are given extra Finnish-language training each day to catch up.[2] It is sobering to realize that Finland has succeeded so well, despite starting with few resources and having a language that is difficult for immigrants to learn. With such a strong commitment to the individual child, Finland now has a superb educational and social system. It features a pregnancy allowance for forty working days, a parental allowance for 320 working days that is divided equally between both parents, plus a substantial child allowance up to the age of seventeen. This contrasts markedly with the US approach of intense competition between schools for funding, which leads to increased numbers of tests to evaluate teachers and students.

Today, political systems in the United States, Britain, and elsewhere have been inundated by misinformation. With the Russian invasion of Ukraine, resistance to misinformation has become critically important. Finnish people are the most resilient to disinformation among all Europeans, as measured by the Media Literacy Index compiled by the Open Society Institute in Sofia, Bulgaria. The Finns attribute this to their education system. In language classes, students are asked to write the same story from a propaganda point of view and from a scientific and factually based point of view. In art, they make an advertisement for a product (soap, for example) from a factually based perspective and another from a marketing perspective. In mathematics, they are shown how statistics can either inform or misinform through manipulation.[3]

In the spring of 1973, I was confronted with girls who were in their formative teenage years; teaching them was an enormous responsibility, and I was keen to do my best. My teaching load was large. I taught the final two years of biology, plus eighth- and ninth-grade general science, with an occasional chemistry lesson if the teacher was absent. With the Finnish system in mind, and remembering the huge impact my geography teacher, Miss Crawford, had had on my own development, I employed her technique of getting the students to ask their own questions and design their own experiments to answer them. I was always aware of the stimulation of asking questions to which I did not know the answer. Such questions galvanized the children to speculate and motivated them to explore books about their topic in the library and ask their parents' opinions at home. I spent all my lunch periods in the laboratory, which quickly became a hub of activity, filled with aquariums, terrariums, and apparatuses and with children of all ages doing their own experiments. Often the older students helped the younger ones. It turned out occasionally that the children slowest to learn in the standard way were the most imaginative and inventive, and it was enormously rewarding to see them gradually become more confident and to observe how their peers began to treat them with more respect. We spent time on expeditions into the nearby woods and fields, and on rainy or snowy days I sometimes showed films, as Miss Crawford had at my boarding school.

Some of the fun in teaching classes of eighth-grade students came from asking the children to design a fictitious plant or animal that could survive in extreme desert or arctic conditions. They were amazed that products of their wildest imaginations did exist in nature, if not exactly, then close, which provided opportunities for discussions about why those differences occurred.

One exceptionally stimulating class for my ninth-grade students was inspired by my reading about a similar exercise for Finnish schools. I asked the students to read four current papers on resource distribution and global population that I had obtained from the United Nations office in Montreal the week before. These papers were simple and clearly written for the layperson. Nevertheless, the children felt very grown up and important reading these official papers. I drew a map of the world on the floor and divided the children into wealthy and poor countries, according to the population and resource statistics mentioned in the papers, and gave them M&M's candies in proportion to the resources. I then asked them to debate two questions: Why did some countries have more resources than others? What could be done about the inequalities in the world? The resource question led to discussions of climatic and geological differences. For the inequality problem, the poor group criticized the wealthy group for being selfish, and the wealthy group argued its reasons for remaining prosperous, such as hard work. The next day, I switched the children; those who had represented the affluent countries now represented the impoverished ones and vice versa. The boot was now on the other foot; both arguments and roles were reversed. The debate became so intense that when the bell rang for the end of class, the students were still arguing, and it ended with the representatives of the wealthy throwing their M&M's at the poor in a fury, saying, "Take them, take them all!" When I went into the staff room at the end of the day, other teachers asked what I had done—the children were still debating throughout their French, mathematics, and English classes! Because the papers I gave them were official United Nations documents available to the public, the children could not understand why, when faced with such obvious discrepancies, governments throughout the world did not cooperate and solve the problem. Haven't we heard this before!

I was curious to know how well my classes did in comparison to those taught in a more traditional way. The answer was that all my students did exceptionally well in their Canadian biology high school diploma, equivalent to advanced-placement (AP) biology in the United States and A level in the United Kingdom. I felt vindicated and had paid a debt to the memory of Miss Crawford and to ideas obtained from the Finnish system. Perhaps the biggest compliment came seven years later. I had driven Nicola from Ann Arbor to Dartmouth College for her second year at university. Thalia's high school had not yet started fall classes, so she came with me. I suggested that we visit our old Montreal haunts and friends on the way back. We were walking through the woods of the Morgan Arboretum at McGill University's Macdonald Campus when a couple of women stopped us. You will not recognize us they said, but we are mothers of two children you taught at ECS. We want to tell you that every child you taught in our daughters' class, whether they became doctors, lawyers, or something else, all became involved in activities related to climate change, conservation, and other ways to reduce inequalities among people and protect the precious biodiversity of our planet. It is hard to think of a higher tribute than to have played a small role in inspiring members of a younger generation. Could we change the trajectory of human influence on our fragile planet by adopting Finland's approach to the education of our children?

Teaching gave me another bonus. With the long summer and winter vacations between school terms, I could get back to research.

15

Three Steps Forward

All our knowledge has its origin in our perceptions.

—LEONARDO DA VINCI

Peter and I returned to the question of speciation that we had discussed in embryonic form many years before. While both of us were interested in the differences among individuals in a population, Peter's approach was primarily ecological and mine was genetic. My interests were how phenotypic and genetic variation are maintained, and how, in association with a changing ecological and behavioral environment, they are translated into morphological change that could under certain circumstances lead to a new species. Our two approaches to the same questions—how and why new species are formed—were thus complementary.

How do you tackle such a large and fundamental question as speciation? We discussed different adaptive radiations (young species all derived from a common ancestor) as suitable systems to study, such as the charr, my original proposal for a PhD. Charles Darwin and Alfred Russel Wallace had both suggested that recently diverged species would be rewarding organisms to study. Darwin wrote, "Those forms which possess in some considerable degree the character of species, but which are so closely similar to some other forms or are so closely linked to them by intermediate gradations, that naturalists do not like to rank them as distinct species, are in several respects the most important to us."[1] Wallace wrote in a letter to Henry Walter Bates, "I should like to take some one family to study thoroughly, principally with a view to the theory of the origin of species. By that means I am strongly of opinion that some definite results might be arrived at."[2]

There are many examples of adaptive radiations, all with their advantages and disadvantages for in-depth study. The Icelandic charr, which I had hoped to study, was one example. I could measure individuals and through serological tests of antibodies to proteins show genetic differences and relatedness, and in an aquarium in the laboratory do behavioral tests. However, a big disadvantage was the difficulty of marking and following individual fish in their natural habitat. There are many other examples of adaptive radiation: Hawaiian honeycreepers, lizards in the Caribbean, and oak trees. In New Zealand, the *Hebe* (or *Veronica*) group in the Plantaginaceae, comprising plants that range in habit from small trees to tiny alpine shrubs, provided another possibility; the 124 species all apparently derived from a single common ancestor in the past five to ten million years.[3]

Peter and I had both read David Lack's book on Darwin's finches.[4] The young radiation of birds in the remote Galápagos archipelago could be highly suitable. There, closely related species occur on several islands that are uninhabited by humans. Thus, any differences and changes we measured over time would be the result of natural causes and not human influence. Furthermore, regular fluctuations in the climate would affect the food supply and hence survival of the finches.

In the Galápagos, the normal wet season extends from approximately January to June; the remainder of the year is dry and cool with the only moisture coming from *garúa,* or sea mist. Superimposed on this bi-seasonal cycle is the El Niño–Southern Oscillation phenomenon (ENSO), which brings torrential rains once or twice a decade followed by a drought lasting from one to two years. In drought years, many birds die of starvation. Who dies and who survives depends upon the food supply and the ability of individuals, in terms of their behavior, body size, and beak size and shape, to exploit it.

Peter was excited because of his interest in competition. If two species of different beak size and shape live on the same island and depend on the same diminishing food resource during a drought, competition between individuals that were most similar would occur, causing the differences between species to be accentuated, a phenomenon termed *character displacement.* Peter had attempted to study this possibility in the rock nuthatches in Iran, but the results were inconclusive because of the logistical difficulties of following individual birds across generations. I have always admired Peter's fresh approach to a problem, his insights that lead directly to the heart of a question, and his tenacity in delving deeply. His comprehension of nature and perceptiveness usually ended up with him tackling research questions that were unique and out of the mainstream at the time.

I was intrigued by the unusually high variation in beak shape of the Large Cactus Finch, *Geospiza conirostris* (now classified as *G. propinqua,* the Genovesa Cactus Finch), on the Galápagos island of Genovesa, mentioned in David Lack's book, and thought that this island would be a good location for my own research. To give an idea of the extent of the variation in morphology, I have used statistical measures of variation either side of the mean (average) of beak measurements. The average coefficient of variation for beak size in continental sparrows and finches is two to five—that is, the standard deviation is 2 to 5 percent of the mean—whereas the coefficient of variation for beak size in *G. conirostris* on Genovesa is in the range of five to seven for different dimensions of the beak. This high variation is unexpected because the population is small and lives on an isolated island in the far northeast of the archipelago. The prevailing winds and currents from the southeast would deflect any dispersing finch westward, deterring possible migrants to the island. In such a remote situation, we would expect the population of finches to become inbred and lose genetic variation. Why were they among the most variable populations in the archipelago? How did this variable population arise? How was so much phenotypic variation maintained? How did this population, which according to Lack was small, survive the intense droughts that periodically hit the Galápagos? Answering these questions could shed light on the processes involved in maintaining the variation that is important for evolution to fuel speciation.

The drawback was the Galápagos archipelago was far away, and it seemed impossibly expensive to carry out such research. The breakthrough and motivation for finding money came when Peter received a letter from Ian Abbott, a young Australian who had recently obtained his PhD. Ian shared Peter's interest in the question of whether competition between species occurs in nature and, if it does, how it could be detected. Having read Lack's book and Peter's papers on competition in mice, Ian was interested in collaborating with Peter on postdoctoral research. His goal, like Peter's, was to test the hypothesis that competition for food caused divergence in beak size in Darwin's finch species. After Peter applied for money elsewhere and failed to get any, McGill gave him $4,000. This was enough for

Ian and his wife, Lynette, a botanist, to spend four months in the Galápagos in 1973 and for Peter to join them for one of those months.

From January to May 1973, Ian and Lynette visited many of the archipelago's islands, caught and measured birds, documented their feeding behavior, and did a detailed assessment of the food available to them by painstakingly identifying all the plants, and counting the seeds, in quadrats. This was invaluable for our research, as they identified the plants that gave us baseline data for food items in the form of seeds of different species and their abundance. Peter spent that April with the Abbotts on Daphne Major and Santa Cruz Islands, participating in the fieldwork. Later that year, Peter and I went there with Nicola and Thalia, ages eight and six, for four weeks over our schools' Christmas break.

We flew the four hundred miles to the Galápagos archipelago in a tiny low-flying military plane from Guayaquil, Ecuador. The mainland had suffered torrential rain for several weeks, and we saw huge clumps of water hyacinths and logs, tangled together, propelled from the mouth of the Guayas River. Some of these vegetation rafts looked like miniature islands, which we estimated to be more than three hundred feet wide, and were miles out in the open sea on their way toward the Galápagos. Later, on beaches on Genovesa, we found several large Balsa trunks that must have originated in South America, as Balsa trees do not grow on the Galápagos, and we suspected they had arrived in this way. Geckos, lizards, snakes, iguanas, and even the ancestor of the giant Galápagos Tortoise, symbol of the islands, may have arrived from the South American mainland aided by such rafts of vegetation. Although giant tortoises are not good swimmers, they can float after being washed into the sea and survive without fresh water and food for as long as six months.[5]

My first impression of the islands was that they were young; the volcanic landscape had signs of recent activity in the form of unvegetated lava flows. Questions immediately came to mind. How old are the islands? Did their formation and dynamism influence the establishment and subsequent radiation of the finches? I later learned that geologists Felipe Orellana-Rovirosa and Mark Richards had examined seamounts east and northeast of the present-day Galápagos archipelago to reveal the presence of past archipelagoes. Islands had emerged over the past twenty million years or more through volcanic activity at the conjunction of the Cocos and Nazca tectonic plates, an area known as the Galápagos Spreading Center. From there, the islands were carried eastward toward the South American mainland on the Nazca Plate, staying above sea level for approximately five million years before subsiding to become seamounts. Thus, archipelagoes formed and subsided on a slow-moving conveyer belt, changing in size, shape, and number of islands over time. An archipelago calculated to have been in place sixteen million years ago was huge, over two and a half times the size of the present archipelago.[6] During the past million years, ten glacial-interglacial cycles in which water was alternatively locked up as ice at the poles and then released caused oscillations in the sea level by at least 425 feet. This resulted in islands fusing and separating through time, a dynamic landscape of ecological changes on which the radiation of finches evolved.

When we arrived on Daphne Major that December, it was the dry season. We found that more than 80 percent of the birds Peter and the Abbotts had measured and banded were still present, whereas those on the larger islands had almost entirely disappeared. Most of those, we thought, had either died or dispersed up into the highlands, where there was moisture and a greater supply of food. Peter decided that, as a small island with large numbers of two species of finches that were resident year-round, Daphne was the place to do an in-depth study.

For this visit we were joined by our good friends Jamie Smith and Yael Lubin, both doing postdoctoral fellowships in Panama. This was a wonderful four weeks. We adults threw ourselves into fieldwork and had discussions on all aspects of the research. Yael's interest was in spiders, and she took us out at night to watch the webs being spun, delighting Nicola and Thalia. The children were in their element, helping us to find birds, exploring, and becoming intrigued by the tameness of the wildlife.

In the summer of 1975, we spent another three months as a family in the Galápagos, mainly on Daphne. In November 1977, we returned for a five-and-a-half-month stay. I had given up school teaching at the end of the 1977 academic year, as we were about to move to the University of Michigan in Ann Arbor, where I would be able to carry out full-time research as a research associate. Since we had taken the girls out of school for a few weeks, we compensated for the lack of schooling, and at lunchtime they did their schoolwork in the shade. When they returned to school, they were ahead of their class in all subjects. Each had written long letters about their experiences to their teachers, their friends, and their grandparents. The teachers were thrilled by their letters and told us we could take them out of school any time, jokingly saying they learned more out of school than in.

Finally, in January 1978, I was able to get to Genovesa. The island is a low, flat shield volcano, approximately five miles in diameter and with a central saltwater lake. In those days without GPS, it was a tricky island to find, thanks to devious currents and fickle winds. Many boats were blown off course and had to return to the central islands. The best navigators were the

FIGURE 10. Genovesa Island, Galápagos. Upper: Darwin Bay. Middle: Lagoon. Lower: Arcturus Lake.

experienced fishermen, who constantly tested the speed and direction of the currents with their hands and used direction of the wind and position of the stars to navigate.

We had spent a couple of weeks on Pinta Island with Dolph Schluter, who was doing his PhD research under Peter's supervision on competition between the group of finches on that island. From there we hired a small fishing boat to take us from Pinta to Genovesa, with an overnight stop on Marchena. We had all our research equipment, camping supplies, food, and water to last for four months.

Shortly before we saw Genovesa, the fragrance from *Bursera* and *Croton* flowers carried on sea breezes reached us. Eventually, the island appeared on the horizon, glowing in the early-morning light. The cries of petrels, Swallow-tailed Gulls, tropicbirds, and boobies echoed off the tall cliffs as we anchored in Darwin Bay, a collapsed volcanic caldera. We waded out onto a small coral beach. The tide was high and had filled a lagoon behind the beach that was surrounded by low-growing *Cryptocarpus* plants dotted with nesting frigatebirds. The gular pouches of the courting males vibrated like red balloons against the dark green foliage, attracting the females that were circling in the air above—my introduction to the first of ten years of intensive study on this island.

Genovesa is one of the younger islands; potassium-argon dating by Allan Cox suggests it is around 750,000 years old.[7] In contrast to Daphne, an approximately 20,000-year-old friable tuff cone of compacted volcanic ash, Genovesa is made up of hard, ropy *pahoehoe* lava interspersed with areas of *a'a*, or clinker, lava that sounds like dinner plates clashing together as you walk across it. This jagged surface is broken by deep narrow crevices that become obscured by *Ipomoea* vines after heavy rains, making walking treacherous. A child could quickly plunge into one of the deep crevices. Even the toughest hiking boots wore out within a month or two on Genovesa, and we always ended the field season with duct tape holding the soles to the uppers. The central lake, Arcturus, is less than 6,000 years old. The mineral-saturated water was undrinkable, for us, but full of movement caused by many backswimmers (notonectid bugs) and the larvae of other insects. A small flock of White-cheeked Pintail ducks lived among the mangroves surrounding the lake.

On this rocky island with little soil, plant diversity is low. There are only four species of trees: *Bursera graveolens, Croton scouleri, Cordia lutea,* and *Erythrina velutina,* the last represented by one lone individual. The cactus *Opuntia helleri* is patchily distributed. After the rains, the herbaceous shrubs *Waltheria* and *Chamaesyce* bloom, and annual plants *Ipomoea, Heliotropium, Abutilon,* and *Sida* produce flowers. As on Daphne, all flowers are either yellow or white. Wherever there is a hint of soil, it is occupied by one of two species of *Ipomoea* vines, a sedge, or one of three species of grasses. These are the plants that provide food for the finches, in the form of either seeds, fruits, and pollen or the insects that hide in them or pollinate their flowers. *Opuntia* and *Cordia* bloom before the rains arrive and provide nectar and pollen for the birds, augmenting a dry season diet of seeds and termites and other insects found beneath the bark of trees and in rotting *Opuntia* pads. After the rains arrive on this seasonally dry island, food for birds is abundant. Not only are new fruits and seeds produced, but the plants are host to caterpillars and spiders, which are fed to nestlings and fledglings. The ubiquitous white and yellow flowers made us wonder about their pollinators—the only bee, an endemic carpenter bee, was not present on Genovesa at that time—so we performed a side project. We bagged flower buds to prevent insects from accessing them, and then watched who visited the flowers after removal of the bag. We discovered that finches were major pollinators of both *Opuntia* and *Cordia* flowers.

We used the same technique for studying the birds as we used on Daphne. We caught the birds in mist nets, always careful to take them out immediately, before the sun struck the net. In the cool shade, we weighed them and recorded three beak measurements—length, depth, and width—and wing and tarsus length. Before releasing the birds, we put four bands on their legs, a unique combination of three colored bands coded to a number on the fourth, a metal band. In later years we would also take a drop of blood for DNA analyses. Banding the birds allowed us to identify individuals of known size and beak shape in the field. After banding had been completed, we spent the remainder of the day documenting breeding pairs; counting the number of eggs in each nest and the number of nestlings and fledglings produced; recording songs; and noting down the food items taken by each bird and the amount of food available, tallied by counting fruits on numbered plants and seeds in quadrats on the ground. Our data files look like a register of human births, marriages, and deaths!

We found that the degree of heritable variation in six traits—beak length, depth, and width; body weight; tarsus and wing length—was uniformly high (0.69–0.81), as measured by the association between mid-parental measurements (that is, the average of father and mother measurements) and mid-offspring measurements when fully grown. Zero (0) indicates no heritability; one (1) is 100 percent heritable. These values were similar to those found in *Geospiza fortis* (Medium Ground Finch) and *G. scandens* (Common Cactus Finch) on Daphne.

Results from the feeding observations during the dry season revealed an interesting division in the *G. conirostris* (now *G. propinqua*) population. Blunt-beaked individuals stripped bark off

Bursera trees to obtain insects, often termites, and did the same with old *Opuntia* pads on the ground, from which they obtained Diptera fly larvae. Long-beaked individuals of *G. conirostris* hammered holes in *Opuntia* cactus fruits to obtain seeds, which they cracked open. In the wet season, birds of all beak dimensions ate caterpillars and consumed pollen and nectar from *Opuntia* flowers. By continuing the study for many years, we found that the amount of rainfall influenced the vegetation and the finches' food supply, which caused selection events with evolutionary consequences. The long-beaked *G. conirostris* had a survival advantage in years of abundant cactus-fruit production but suffered when the fruit supply was low, whereas the blunter-beaked birds survived on the supply of insects beneath the bark of trees and in rotting cactus pads during those years.

In 1977, the year before we arrived on Genovesa, there had been an extreme drought, so we were not surprised to find that finches were scarce. Populations of all four Darwin's finches on the island—Gray Warbler-Finch (*Certhidea fusca*), Sharp-beaked Ground Finch (*G. difficilis*, now *G. acutirostris*), Large Cactus Finch (*G. conirostris*, now *G. propinqua*), and Large Ground Finch (*G. magnirostris*)—were low. However, we were surprised to find two types of song sung by *G. conirostris* males. Each male sang only one song type, and remarkably, no two adjacent, territory-holding mated males sang the same song type. In contrast, the distribution of territories of males without mates was random with respect to song type. The difference between the territories of mated and unmated males indicated to us that female choice of mates was involved. The following year, this alternating pattern of territories disappeared. The puzzling, apparently transient pattern of territories motivated us to investigate in detail song and other factors involved in species discrimination, mate choice, and territory establishment.

We found, through response to song playback, that individual *G. conirostris* males recognized both song types as *conirostris* songs and could discriminate between their own species' songs and the songs of the other two ground finch species on the island. Likewise, our mount experiments using stuffed museum specimens revealed that individuals could discriminate between their own species and another species by appearance alone, in the absence of song, but did not discriminate between blunt- and pointed-beaked individuals of *G. conirostris*. These experiments were first done with Laurene Ratcliffe, who was studying for a PhD supervised by Peter and doing her fieldwork mainly on Daphne. She joined us for two weeks on Genovesa and continued her song playback and mount experiments, using the same experimental design Peter had used in Iran with the nuthatches. The mount experiments involved stuffed museum specimens of females of two species, each one attached to opposite ends of a pole. This apparatus was then taken into the territory of a living male at the height of the breeding season and presented to him to test whether he could discriminate between a female of his own species and a female of another species. The results were striking: males vigorously courted the female of their own species, sometimes to the point of copulation, and ignored the female of the other species. All song and mount experiments were carried out in many different territories and with controls (non-Darwin finches). It turned out that both factors, song and morphology, but particularly song, are important in species discrimination.

Darwin's finches are among the species that learn their song early in life, and once learned, it is retained unaltered for life. Other birds, such as thrashers, can learn and extemporize throughout life, while approximately 50 percent of bird species (e.g., ducks, geese, and swans) cannot learn songs but have genetically programmed sounds. Robert Bowman's research using captive birds in the 1950s showed that all Galápagos finch species learn their song approximately ten to forty days after hatching. Thereafter, their song remains unaltered for life, as shown by our yearly recordings. This sensitive period for learning spans the time when the young birds are in the nest during their last few days before fledging and when they are out of the nest still being fed by their parents. All this time, their father is singing. In this way, the young males learn their father's species' song and associate song with species appearance. Song pedigrees can be traced from grandfather to father to son, down the generations like last names in our Western society. This is possible to detect because of small individual variations within the species' song. Females do not sing, but our observations through pairing showed that both males and females mated according to the species' song learned early in life. Such a phenomenon, first documented by Konrad Lorenz, is termed *sexual imprinting* and results in a premating barrier to interbreeding between species.

However, this barrier is based on learning and as such is vulnerable to leakage if a young bird hears and learns the song of another species during its short sensitive period early in life, as we observed on rare occasions. This may occur when a father dies and leaves the mother to raise the offspring. The young bird then hears the song of a neighboring male. If the neighbor is a member of another species, it learns that species' song. This leads the bird, when adult, to attempt mating with a member of that species. Learning another species' song can also occur after a nest is taken over by a member of another species and an egg of the original occupant is left behind. These behaviors—imprinting on paternal song during a short sensitive period early in life, lack of singing in females, retention of the song throughout life, and pairing according to species-specific song—we later found occur without exception in all species of Darwin's ground finches that we studied.

The witnessing of rare hybridization through this process was unexpected because we had been told by David Lack in person that there was no evidence of finch hybridization. He had looked for it in 1940 but observed none, and as far as anyone knew at that time, no one had ever observed Darwin's finch hybrids in nature. Thus, we were unprepared for our own detailed observations of *G. conirostris* that revealed rare hybridization with the two resident ground finch species, *G. difficilis* and *G. magnirostris*, on Genovesa. Backcrossing—that is, the breeding of hybrids with a member of one or other of the parental species that produced them—causes a trickle of genes to pass from one species to another, thus increasing the genetic variation. We calculated that the genetic input from *G. magnirostris* and *G. difficilis* through interbreeding and backcrossing increased the coefficient of variation by 20 percent. This was a wonderful example of the value of my father's principle to "always follow your exceptions."

Hybrids survived well and bred until a very severe drought in 1986, when all known hybrids died due to the lack of seeds and easily available insects. We concluded that hybridization followed by introgression (the transfer of genetic material between species) is the crucial factor contributing to the high level of variation in *G. conirostris*. It is sufficiently frequent to counteract the loss of genetic variation by random drift but not so frequent as to prevent selection from favoring certain combinations of alleles important for adaptation. The reason the populations of the different species do not merge, we found, is they have different diets in the dry season, and these differences are accentuated during extreme droughts. Years later, introgression of genes was confirmed in greater detail through studies of the genomes of these species by our colleague Leif Andersson and his laboratory group using the stored blood samples we had collected.

Another factor that turned out to be important in both the setting up of territories and the choice of mates within a species is delayed plumage maturation. All ground finch males gradually acquire fully black plumage over several years. The speed at which they acquire fully black plumage varies among species and individuals within species. For example, *G. fortis* on Daphne, molting into gradually darker plumage each year, takes between four and six years to develop fully black plumage. First the head and then the neck become black, and finally the whole body follows. *G. conirostris* changes faster, taking only two to three years to turn all black. A male in partially black plumage signals its youth and immaturity and is subordinate to older fully black males, we found through observations and experiments. Breeders with immature plumage establish their territories in the interstitial places between territories of older, fully black males. If they obtain a female, it is usually a young bird, also breeding for the first time, or an older female whose mate has died. Once males have bred successfully, they retain their territory for life. The life span can be as long as seventeen years. We published the results of our ten-year intensive study in several papers and in a book entitled *Evolutionary Dynamics of a Natural Population: The Large Cactus Finch of the Galápagos*, which won the Wildlife Society Wildlife Publications Award in 1991.[8]

Our findings show that the exchange of genes through hybridization can be crucial for the conservation of many species that are trapped or have evolved in isolation and are in danger of extinction through inbreeding and loss of variation. In such cases, it is important to conserve more than just one focal species. Without *G. magnirostris* and *G. difficilis*, Genovesa's small *G. conirostris* population would have become inbred and possibly gone extinct during one of the many droughts that impact the island.

I did most of my Genovesa research during the six years we were based at the University of Michigan in Ann Arbor, spending three to four months each year in the Galápagos. Dolph Schluter and Trevor Price were Peter's graduate students at the University of Michigan. Laurene Ratcliffe and her husband, Peter Boag, were finishing writing their PhD theses at McGill and visited us from time to time. Lisle Gibbs did his PhD research on Daphne after Trevor had finished. Margaret Kinnaird and Bob Curry joined the group to study the Galápagos Mockingbird, basing their work on research that had been started by Nicola. Collectively, while in Ann Arbor, we held weekly lab meetings, which attracted other professors, graduate students, and postdocs. It was in this energizing, spirited environment that I completed my *G. conirostris* research, publishing papers in peer-reviewed journals, though I was still without a specific plan for a PhD degree.

The final steps to a PhD degree were as unpredictable as any along my erratic career path. I was asked to give talks about my *G. conirostris* research in Sweden, first at the University of Gothenburg and then at Uppsala University. At the Uppsala talk, Professor Staffan Ulfstrand introduced me as Dr. Grant. I told him I was not Dr. Grant as I did not have a PhD. At a party in the evening, he asked whether I would like to do a PhD degree at Uppsala. "If you are interested," he said, "I will be your supervisor." The next day I went to him and asked if he was serious. He said, "Yes. For the thesis, put all the papers you have already published in peer-reviewed journals together with the two you are writing once they are finished. You will need to take a couple of exams and put in a semester of residence. If you are successful, you will host a lunch for all those who attend your public oral defense and a party at night for the whole department in celebration!" This was feasible because it would coincide with another of Peter's sabbatical leaves.

I followed the ancient Swedish tradition of nailing the thesis to the central building of the university and advertising the thesis defense on the local radio and television stations. The department arranged for an opponent and jury from other universities. André Dhondt from Belgium, who was studying bird populations in fragmented habitats, their differences, behaviors, and now disease transmission, was my opponent. I knew of him and his research but had never met him. I was told firmly by Staffan not to answer questions with *yes* or *no*, but to turn them into an interesting discussion. I could not have had a better opponent; he was knowledgeable, insightful, probing, and polite. We created, I hope, an enjoyable, intellectually stimulating, if (for me) nerve-racking, discussion. At times, it was like sparring with my father! My fellow graduate students had put little notes of encouragement behind a row of flowers on the desk in front of me. The lunch was held in the department's museum, and the party at night featured clever skits at my expense from my peers and music from a university band made up of fellow graduate students. We provided a splendid dinner, all organized with the help of our friends Gunilla Rosenqvist and Ola Jennersten. Peter and I walked home at three a.m. smiling to the moon. Years later, I was an opponent at Uppsala and decided it was much easier to be the degree candidate than the opponent. My admiration for André was considerably enhanced. He had done a magnificent job.

FIGURE 11. Upper: With Thalia and Peter, looking at a map of Galápagos, Ann Arbor, Michigan, 1982 (photo by Jim Jagdfeld). Lower: My PhD celebration at Uppsala University, Sweden, with Staffan Ulfstrand (right) and Peter, who received an honorary degree, 1985.

Later, in May of 1986, we attended the conferment ceremony at Uppsala University, in which I received my PhD and Peter received an honorary degree, while my parents watched from the audience. Today we each wear the Uppsala doctorate ring, a gold-embossed laurel wreath, a treasured emblem of our scientific marriage, but the ring that lies below it and nearest to my heart is our wedding ring.

16

Daphne Research

> Look deep into nature and then you will understand everything better.
>
> —ATTRIBUTED TO ALBERT EINSTEIN

The research on Daphne had continued uninterruptedly since 1973, done by a series of graduate students—Peter Boag, Laurene Ratcliffe, Trevor Price, and Lisle Gibbs—and us, visiting for several weeks each year. In 1991, Peter and I decided to give our full, joint attention to the problem of understanding finch evolution on Daphne. Research on Genovesa continued for a few years with the aid of assistants David Anderson, Tom Will, Michael Wells, Anne Heisey, and Jim Waltman, and graduate student Bob Curry, who did his PhD on mockingbirds. In the meantime, Peter and I moved from Montreal to the University of Michigan, in Ann Arbor, in 1978, and then, in 1986, to Princeton University, in New Jersey.

Daphne Island is more suitable for our research than Genovesa because it is smaller, which makes it possible to band, measure, and follow the fates of every individual finch on the island. On Daphne, the population of *Geospiza fortis* (Medium

Ground Finch) ranges from over a thousand individuals during wet years to around a hundred in drought years, and that of *G. scandens* (Common Cactus Finch) from over seven hundred to ninety. There is an amazingly high density of finches for such a small island.

Living on this island for several months every year for forty years, we feel we know almost every rock but become acutely aware of the subtle and not-so-subtle changes that occur each day—how changes of the moon and direction of the wind influence the height of tides, and how the appearance of cumulus clouds over Santa Cruz presages imminent heavy rain. Yearly changes are noteworthy, not only the differences between El Niño and La Niña events but the less obvious differences—masses of jellyfish one year, an albatross and a penguin another year. Daily we see Marine Iguanas and sea lions invade our kitchen area, yet once in a while we have a hybrid iguana and an errant fur seal. Our kitchen, a cave on the outer slope on the cliff, overlooked the sea at a point so deep that boats cannot anchor yet so clear you can make out sharks resting on the rocks many feet below. It was like being perched on the side of an overcrowded aquarium with colorful reef fishes passing back and forth and the occasional pair of mating turtles swimming by. Pelicans shiver on the cliff edge before plunging in after fish. Schools of young tuna occasionally dart past chased by sharks. This attracts a feeding frenzy of boobies, petrels, and shearwaters, which dive after the fish from above. Escaping predation from above and below causes the schools to wheel and turn in unison, for it is only by being one fish in thousands that there is hope of not being caught and eaten. Swallow-tailed Gulls, surely the most graceful of their kind in the world, resembling terns more than other gulls, nested around us, feeding their chicks regurgitated squid. Each afternoon the tropicbirds would leave their shallow crack in the rocks to form screaming parties that swoop and swerve around the island before going off to their feeding grounds, far out to sea. No wonder we looked forward to our yearly months spent on this island.

Daphne is the tip of a volcano rising above the surface of the sea, technically a pyroclastic tuff cone, a mere half mile in diameter and four hundred feet high, with a steep-sided double crater in the center. The finches breed in cactus bushes on the inner and outer slopes, and since they go everywhere, so must we.

Forty years of intensive fieldwork on Daphne by us and the graduate students, postdoctoral fellows, and assistants gave us answers to our questions about evolutionary responses to natural-selection events and three possible methods of speciation. Our discoveries depended on intensive fieldwork, measurements of individuals, and as I described for Genovesa (chapter 15), a registry of the births, matings, offspring production, and deaths of all individuals. In addition, we had detailed counts and records of changes of food availability. However, in the first few years, we knew nothing about the underlying genetics that made possible the changes we witnessed.

While still in Montreal, with genetics in mind, I had learned how to karyotype my own blood cells to look at the complete set of chromosomes in a cell by staining and imaging. My goal then was to get some information about the genetic basis from chromosomes in the pulp at the base of a bird feather. I thought it might be possible to determine the number of chromosomes for each finch species and, if possible, any obvious deletions or duplications in the chromosomes. Everything went well in that I could see the large chromosomes, but I could not reliably distinguish the microchromosomes, of which there were many, from the minute granules of melanin, so I had to give this up.

FIGURE 12. Upper left: Daphne Island, Galápagos, from the sea. Upper right: Daphne's crater from the air. Lower left: Cave on Daphne used as a kitchen. Lower right: Bedroom tent on Daphne. Far right: Easiest way to climb onto Daphne (photo by Martin Wikelski).

FIGURE 13. Upper left: Standing at mist net in crater on Daphne Island, Galápagos. Upper right: In banding cave on Daphne. Lower left: Dry conditions on Daphne. Lower right: Wet conditions in same place.

Then a breakthrough came in 1984. Alec Jeffreys at the University of Leicester developed a microsatellite technique known as genetic fingerprinting. Microsatellites are small, noncoding repetitive units of DNA that are unique to an individual and can therefore be used to analyze pedigrees and characterize a population of birds. Studying them required collecting small drops of blood for DNA analyses. We were told that the only way to do this was to take a cylinder of liquid nitrogen into the field and, with a capillary tube, take a drop of blood and fast freeze it in a buffer solution. This we did in collaboration with Peter Boag and his student Gilles Seutin in 1988. Gilles and I tried landing on Española Island in a small boat with the cylinder between us. The sea was rough, and as we neared the shore, a heavy roller caught the bow of the boat, sending us and the cylinder flying into the sea. As I surfaced, Gilles was shouting, "*La bomba, la bomba.*" Together we grabbed the bobbing cylinder, fortunately still upright, and between the waves managed to run with it to the top of a steep sandy shore. Gilles calmly remarked that, if it had tipped, we would both be dead! I was furious. "Gilles, there must be a better way to collect blood; this is far too dangerous." Gilles insisted it was the only way. We argued in English, French, and Spanish.

Necessity is the mother of invention. As soon as Peter and I returned to Princeton, we obtained the help of molecular geneticist Marty Kreitman to try out several techniques of collecting and storing DNA from chicken blood. Marty wisely said, it should work if the blood were preserved in ethylenediaminetetraacetic acid (EDTA), a binding agent, and never exposed to air. This involved putting a drop of EDTA on the brachial vein of a chicken, pricking through it with a sterile needle, and obtaining the required tiny drop of blood on filter paper previously moistened with EDTA. This worked magnificently. After air-drying, the sample could be stored in a bottle of Drierite, and no refrigeration was necessary. Another advantage to this method was that we could write the bird's identity (from its bands), the date, and the place of capture on the same piece of filter paper with the blood sample. In Princeton, we now have thousands of such samples collected year by year since 1988, kept in a freezer set at minus eighty degrees Celsius. The beauty of this technique is that small bits can be cut off the initial spot on the filter paper and analyzed as new genetic techniques become available. Birds have nucleated red blood cells, so only a tiny amount of blood is needed for each analysis. Today we send pieces of these small drops of blood to Leif Andersson's laboratory in Uppsala, where they have been used to sequence the complete genome of all Darwin's finches.

Before the technique of full genome analysis became available, we used microsatellites to genetically identify each individual and confirm pedigrees, matching them with our field data. However, we wanted much more. We wanted to know which genes are involved in the development of different beak sizes. Variation among the Darwin's finch species in beak size and shape is obvious even at the time they hatch from eggs, as we knew from measurements of hatchlings. This told us that to find the genetic pathways involved in beak development, we needed to work with developmental geneticists. Our first collaboration was with Cliff Tabin, a dream come true. I had read many of his papers and greatly admired his research, but I never thought he would be interested in our fieldwork and did not dare to contact him, as he was head of a large medical department at Harvard University and was surely much too busy. One morning, the phone rang when I was in Peter's office. Peter held his hand over the voice piece and said, "Cliff Tabin would like to know if we would be interested in collaborating with him." "WOW! SAY YES," I mouthed back. The next day Cliff drove to Princeton, and this was the beginning of a wonderfully productive collaboration with him, his postdoctoral student Arhat Abzhanov, and later with Arhat's graduate student Ricardo Mallarino, who is now an assistant professor at Princeton.

Cliff, Arhat, and Ricardo's goal was to examine the genes involved in the embryological stages at which differences in beak shape between species first appeared. We helped them to obtain permits to collect the second egg out of a few nests of several species. We decided not to take the first egg, for fear it would cause the parents to desert the nest, but we knew that removing the second egg would not be a problem. The team incubated the eggs in order to follow the expression of genes through development. Initially, Arhat and Cliff found two major signaling molecules—that is, molecules that transmit information between cells. The first gene, *Bmp4*, was expressed in the mesenchyme (embryonic connective tissue that gives rise to cartilage and bone) of the upper beak and was associated with deep and broad beak morphology. It was most strongly expressed in the Large Ground Finch, *Geospiza magnirostris*. Two years later, they discovered that calmodulin (*CaM*), a molecule involved in mediating Ca^{2+} (calcium-ion) signaling between cells, is expressed at high levels in the long and pointed beaks of the Common Cactus Finch, *G. scandens*. Interestingly *Bmp4* and *CaM* are expressed independently in the developing cartilage, so beak shape would be altered if the expression of one was changed relative to the other in amount or timing. A few years later, Ricardo found that when cartilage was being condensed into bone, three other genes, *TGFβIIr*, *β-catenin*, and *Dickkopf-3*, were expressed in the developing beaks. The combined results showed how this sequence of development involved regulating molecules that could be independently modified to produce multidimensional shifts in beak size and shape. This was an exciting advance in our understanding of how differences in beak shape arise in early embryonic development. We published the findings in research papers and our long-term study in the field up to this point in a book entitled *40 Years of Evolution*.[1]

However, this research still did not give us direct insight into the genetic pathways involved in the selection events we had studied. To pursue this topic and others related to it, we collaborated with Leif Andersson and his colleagues at Uppsala University in Sweden. Leif had made groundbreaking discoveries of the genetic basis of phenotypic variation using domestic animals and was interested in extending his research further into natural populations. It was through this collaboration, combined with the forty years of intensive fieldwork on Daphne by us, graduate students, postdoctoral fellows, and assistants, that gave us insight into the genomic underpinning of three possible methods of speciation: allopatric speciation, fusion through introgression, and homoploid hybrid speciation. I will describe our research into each in turn.

Allopatric Speciation with Character Displacement

David Lack, with reference to Darwin, proposed that the radiation began when finch-like birds arrived on one of the Galápagos Islands from the adjacent South American mainland and underwent evolutionary change as they adapted to the new ecological conditions.[2] He argued that when the population increased, some individuals would fly to another island, where they would encounter different ecological conditions and the process of adaptation would occur again. This would happen repeatedly on different islands, resulting in divergence in beak and body size of isolated populations. Eventually, the two diverged populations would come together in secondary contact on an island, where they would compete for a shared food resource. Those individuals that overlapped the most would be eliminated, causing the two populations to become even more separated. Darwin had referred to this process as the "principle of divergence of character." In today's jargon it is called "character displacement." Peter had explored these ideas of character displacement in his studies of nuthatches in Iran, Greece, and Yugoslavia.

On Daphne, it was possible to track populations over time. A severe drought in 1977 resulted in an 80 percent mortality among the *G. fortis* population. Small birds suffered to the greatest extent, after they had depleted the supply of small and soft seeds. The survivors had large beaks that they used to crack the large, hard, woody fruits of *Tribulus* to reach the seeds inside, the only food available during the latter part of the drought. When the rains returned the following year, the large-beaked survivors bred and produced offspring whose beaks were also large, because size is strongly inherited genetically.

The next major drought occurred from 2003 to 2005 and resulted in the death of almost 90 percent of the *G. fortis* (Medium Ground Finch) population. As before, the small soft seeds were consumed first, leaving a dwindling supply of hard *Tribulus* fruits. But circumstances in these two drought periods differed in one important respect. In 2004, unlike in 1977, there were over two hundred *G. magnirostris* (Large Ground Finch) individuals on the island. *G. magnirostris* had arrived during the massive El Niño of 1983, and three males and two females had stayed to breed. A few more immigrants arrived in the following years, gradually building up a substantial breeding population, which remained on the island. *G. magnirostris* is a sister species of *G. fortis* and is twice as large, very aggressive, and well equipped to crack the *Tribulus* fruits and consume the seeds. These *G. magnirostris* individuals outcompeted the large members of the *G. fortis* population for the *Tribulus* fruits, and most large *G. fortis* individuals died. Small *G. fortis* individuals were also dying for lack of small seeds, but a few altered their behavior and followed *G. magnirostris* around, snatching the little seeds that flew out of the locules when *G. magnirostris* cracked open the *Tribulus* fruits. When the rains returned, the few remaining small *G. fortis* individuals bred with each other and produced offspring that were small, due to the high heritability (inheritance) of body size. Thus, *G. fortis* diverged from *G. magnirostris* in body size and beak dimensions. We had witnessed character displacement in action! This was an amazing and fortuitous result, a perfect example of character displacement occurring in real time in a vertebrate population under natural conditions for reasons we understood. Peter's long exploration of competition since graduate days had been justified.

What were the underlying genetic factors? Fortunately, we had blood samples across the years from known and measured birds to address this question. Leif and his graduate student Sangeet Lamichhaney took fragments from our samples, sequenced the DNA, and compared the DNA from all the finch species. They found an important gene that regulates the expression of other genes.[3] It is a transcription facilitating factor, *HMGA2,* which codes for a protein involved in transcribing DNA into RNA and then into proteins. This gene is also found in humans, in which it is known to affect height. In the finches, it comes in two forms: one, which we named *HMGA2L,* is frequent in large species, and the other, *HMGA2S,* is common in small species. Both forms were present in the *G. fortis* population on Daphne. By comparing our blood specimens from finches before and after the selection event of 2004, they found there was a shift in the frequency at this locus from a predominance of *HMGA2L* to *HMGA2S,* which statistically explained 30 percent of the shift in beak size. The selection coefficient at this one locus, 0.59, was very high, meaning that selection was exceptionally strong. Initially, we thought that the response would involve many genes, each with a small effect, in agreement with general theory, so we were astonished to learn that one transcription factor alone could have such a large effect on selection at the genetic level.[4] Of course, these transcription factors would themselves be modified by many small genetic factors.

These findings—divergence of a species adapting to an altered environment, colonization by a competitor leading to character displacement, and identification of a single gene with a disproportionate influence on one of the traits, beak size—

built on Darwin's and Lack's preliminary speculations in ways that must have been unimaginable to either of them. It was not anticipated by us when we began our study on Daphne. These examples of natural selection with an evolutionary response along with character displacement support Darwin's model of speciation, now known as allopatric speciation.[5]

Speciation by Fusion Through Introgression

The second process of speciation involves fusion of two species. In principle, a species may change so much through hybridization that it transforms into another species. On Daphne, *G. fortis* and *G. scandens* occasionally interbreed, and *G. fortis* (but not *G. scandens*) occasionally breeds with the rare immigrant *G. fuliginosa* (Small Ground Finch). Hybridization occurs for the same reason as I described for Genovesa (chapter 15). That is, a bird learns the song of another species during its short sensitive period for song learning early in life, and when it becomes an adult, it mates with an individual according to the species' song it learned. Only by meticulously tracking individuals and recording their songs and mating patterns were we able to pick up the rare 1–2 percent of hybrids produced each breeding season. On Daphne, we followed these few hybrids, but none survived to breed during our first ten years of study. We thought this could possibly be attributable to the incompatibility of the genomes of the two species, but there also was very little food on Daphne during those ten dry seasons, and many young birds were dying. Starvation, not genetic incompatibility, turned out to be the case. The enormous abundance of rain in the 1983 El Niño event brought a wealth of small seeds. With this now plentiful supply of food, hybrids survived to breed, and they did so according to the species' song they had learned early in life. In this manner, via the few hybrids, a trickle of genes flowed intermittently between *G. fortis* and *G. scandens*. Although gene flow occurred both ways, the majority was from *G. fortis* into *G. scandens*. The reason for this predominantly unidirectional gene flow was that the *G. scandens* sex ratio was highly skewed in favor of males during these years. Not having females of their own species available, a few *G. scandens* males bred with hybrid females. Hybrid males were outcompeted for territories and for females by the larger *G. scandens* males. Over the next thirty years, the trickle of *G. fortis* genes into the *G. scandens* population caused the average beak shape of the *G. scandens* population to became blunter and more *G. fortis*–like. Accordingly, the two species began to converge genetically and morphologically, and *G. scandens* on Daphne became increasingly different from *G. scandens* on other islands—a small step in the process of speciation, and showing a second method of speciation, through the fusion of two species.

Once again, we wondered what genetic factors were responsible for the changes we observed in beak shape. With our blood samples from each bird, several thousand in total, Leif and Sangeet, with the aid of Carl-Johan Rubin and others in their laboratory, compared two sets of Darwin's finch species, those with blunt beaks and those with pointed beaks, in search of genetic differences. They found several genes, and the most prominent was a transcription factor, *ALX1*. We were excited about this because this gene is known in humans to be involved in cranial facial development; a mutation in *ALX1* is largely responsible for cleft palate. In the finches, it comes in two forms, which we named *ALX1B* (blunt) and *ALX1P* (pointed). *G. fortis* on Daphne had both but mostly the blunt form; *G. scandens* had the pointed form. Leif and Sangeet were able to track *ALX1B* moving from *G. fortis* into *G. scandens* with our blood samples taken before and after introgression.[6]

By studying whole genomes, our colleagues found that not only were genes on the autosomal chromosomes transferred from *G. fortis* to *G. scandens* but the female-inherited mitochondria were transferred as well. This fitted with our observations in the field. The Z sex chromosome, as anticipated, was left behind in the transfer. (In birds, the two sex chromosomes are labeled Z and W, with females being the heterogametic [ZW] sex.) We anticipated this pattern because of the skewed sex ratio in *G. scandens* and the breeding of hybrid females with *G. scandens* males. In the ZW system in birds and butterflies, unequal chromosome exchange between two interbreeding species is usually interpreted as an indicator of genetic incompatibility; but in this case, we knew that social incompatibility arising from the skewed *G. scandens* sex ratio could explain our results. An intriguing possibility is that social incompatibility could be an initial first step in the formation of genetic incompatibility between species.[7]

We knew from our field observations that the occasional immigrant *G. fuliginosa* that stayed to breed on Daphne bred solely with *G. fortis*. Although *G. fuliginosa* has never been observed breeding with *G. scandens* on Daphne, genes from *G. fuliginosa* do enter the *G. scandens* population via the conduit of *G. fortis*. Interestingly, *G. fortis*, *G. scandens*, and the derived dihybrids and trihybrids experienced approximately equal fitness in terms of survival.[8] The net result of fusion is a population of *G. scandens* that has deviated genetically and morphologically from other populations of this species on other islands.

More recently, Erik Enbody, Leif's postdoctoral fellow, with Carl-Johan Rubin and others from the Uppsala laboratory, have used the finch blood samples to identify six genetic loci of large effect that explain 45 percent of the variation of beak size and shape in the finches. At these loci, multiple genes are clustered together and then transferred as a group between species through hybridization and introgression and subjected to natural selection.[9]

These results are especially relevant to young adaptive radiations in which species differ principally in ecology and behavior. A genome created by hybridization among two or three species is genetically and phenotypically highly variable and this enhances the potential for evolutionary change, as we had suggested in the Genovesa study.

Homoploid Hybrid Speciation

Interbreeding of species may do more than just influence the species' future evolution; through mixing, it can yield a third species evolving along a new trajectory. Examples in plants often involved whole genome duplications resulting in an organism with additional sets of chromosomes. Here we witnessed the formation of a new lineage with no change in chromosome number, a phenomenon termed homoploid hybrid speciation. We gained this insight in a most unexpected way. In 1981, a large bird arrived on Daphne at a time when almost all finches on the island were banded. Trevor Price, the PhD student on the island at the time, caught the newcomer and banded it with the number 5110. The bird looked superficially like a *G. fortis* (Medium Ground Finch), but it was huge, weighing 28.5 grams, whereas an average *G. fortis* weighs approximately 17 grams. It was a young bird in immature plumage. The following year it molted into black plumage and started to sing. Its song was unique; we had never heard anything like it on Daphne or indeed on any other island. The outstanding question was, where had the bird come from? Peter and I caught the bird and took a small blood sample. Once again,

Leif and Sangeet gave us the answer by comparing the genome of 5110 with genomes of all other species throughout the archipelago. They discovered 5110 was not a *G. fortis* hybrid from Santa Cruz, as we had supposed, but a *G. conirostris* (Large Cactus Finch) from the island of Española.[10] Both Peter and I had written independently in our notebooks that 5110 looked like an Española *G. conirostris,* but neither of us thought that was possible because of that island's distance in the far south of the archipelago.

Because of its size, 5110 managed to acquire and defend one of the best territories on the island, but it could not get a mate. After all, from the point of view of a female *G. fortis,* its beak was the wrong shape, and it sang a weird song. After trying to obtain a female for almost two years, it was finally successful at the very end of the second year. A female *G. fortis* had lost her mate and moved into 5110's large territory and mated with him. During the next two breeding seasons, another female *G. fortis* bred with him. We measured, banded, and took a drop of blood from all their offspring. We were curious to find out how the offspring would select their own mates. One bred with a *G. fortis,* but that lineage did not survive. All the other offspring bred with each other. The severe drought of 2004, which caused 90 percent of finch deaths, eliminated all but two of these big birds. The survivors were a brother and sister. When the rains returned in 2005, the brother and sister bred with each other. They produced twenty-six offspring, all but nine of whom survived to breed. A son bred with his mother, a daughter with her father, and the rest of the siblings paired with each other to produce the next generation. All subsequent generations continued this intense inbreeding. In this way, a new lineage was formed; we called it the "Big Bird" lineage because its members were larger than *G. fortis,* its nearest relative.

Interestingly, this lineage was not intermediate between Daphne *G. fortis* and Española *G. conirostris*; there had been an allometric shift, producing birds with relatively large beaks for their body size. They fed mainly on *Tribulus* seeds in the dry season. A comparison of the cracking times of *Tribulus* fruits by the Big Bird lineage and the larger *G. magnirostris* (Large Ground Finch) revealed that the new lineage was just as efficient at dealing with these foods as *G. magnirostris*, but because it was significantly smaller, it required fewer of the fruits, a possible advantage during droughts. The genetic interactions producing this allometric shift (change in proportions) is being unraveled but likely involves *HMGA2* and *ALX1* acting independently because of a recombination hot spot on the chromosome between the two loci.

In all respects, this Big Bird lineage is behaving like a separate species. Males sing a song that differs from the songs of other species on the island—the same as the original song of 5110, which was passed down from father to son in successive generations. They hold large territories, which they defend against each other. Their territories overlap smaller territories held by *G. fortis*, *G. scandens*, and *G. magnirostris*, who they ignore. This unusual Big Bird lineage may survive and prosper, or it may eventually die out through inbreeding depression. It is also possible that occasional introgression from another species may counteract the loss of genetic variation that comes from inbreeding. Regardless of whether the Big Bird lineage succeeds or succumbs, its existence gives us insight into how a new species can be formed.

When we first started to publish the results of our studies, hybridization was not considered to be an important factor in animal speciation. This widespread opinion among the scientific community was mainly attributable to preoccupation with the final stages of speciation leading to complete genetic incompatibility between two divergent lineages. In contrast, Peter and I were interested in the early stages of speciation—how and why populations diverge. In the past thirty years there has been a proliferation of examples of introgression as a source of variation for adaptation to new or changing environments. There are now examples from all taxa, ranging from bacteria with horizontal gene flow to hybridization in plants and animals, including our own early evolution. Denisovans and Neanderthals bred with our ancestors, and we still carry the genetic signatures.[11] For example, a genetic region associated with susceptibility to severe Covid-19 and another associated with protection against severe Covid-19 in present-day humans are both inherited from Neanderthals.[12] We even find examples in viruses. Ruopeng Xie and co-researchers found that a new pathogenic H5N1 flu strain that infects poultry, wild birds, minks, and marine mammals is the result of two viruses, LPA1 and H5N1, infecting the same cell. The genomes of both LPA1 and H5N1 are divided into eight segments, which recombine when confined within a cell to form new combinations, one of which is the highly pathogenic strain capable of infecting mammals as well as birds.[13]

These findings of introgression and their consequences have attracted attention from researchers of human genetics. Mel Greaves and Carlo C. Marley envisage cancer-clone evolution taking place in tissue ecosystem habitats in the way finch populations evolved in different island ecosystems on the Galápagos.[14] Think of a cancer clone in the colon as living and dividing in a very different ecosystem from a cancer clone in the lung. It enters the bloodstream, travels to other organs, and may settle there and exchange genes with a different cancer clone. This process can involve retrotransposons (a genetic element with the ability to copy and paste itself into different parts of the genome by converting RNA back into DNA), as shown by Kathleen Burns, who found that an introgressed retrotransposon through horizontal gene transfer became associated with a tumor suppressor gene and destroyed its function.[15]

Two general messages came out of our research. The first message is that genomic data considerably enhance understanding from field studies. However, reciprocally, a reliable interpretation of genomic data requires a deep understanding of ecology, evolution, and behavior in natural conditions. The second message is that to maintain biodiversity on our fragile planet, we must keep both environments and populations capable of further, natural change. As we have demonstrated, one way this can be achieved is by maintaining the full complement of related species in their natural habitats.

When I think back to my first interest in Icelandic charr, I am struck by the similarities with and differences from Darwin's finches. Both occur in isolated locations, and in both finches and fishes, early imprinting influences future mating behavior, resulting in the formation of a pre-mating barrier between species or morphs. Young finches imprint on song and appearance, learned from their parents during a short sensitive period early in life. Young charr, like other salmonids, imprint on the habitat of their hatching area and their spawning season, and this determines the place and time of their adult breeding. The major difference between finches and fish is in their mobility. Finches occasionally fly to other islands occupied by other species and breed there, resulting in occasional hybridization with the resident species; genes thus flow among the islands. Fish that are trapped in lakes are less mobile, and this seems to me to be a factor in their frequent diversification within a single lake. Charr in Iceland's Lakes Galtaból and Svínavatn are examples of divergence into a benthic (bottom-living) and pelagic (open-water) form within one lake. They are more similar to each other genetically than they are to comparable morphs in other lakes. Lake Thingvallavatn has four resident morphs: small benthic, large benthic, planktivorous (plankton-feeding), and piscivorous (fish-feeding). Here there is evidence of divergence both within and between lakes. The possible influx of charr from another source could be from either another lake or the sea because of a recent history of volcanic activity in the area, which constantly modifies the topography.[16] Another important difference is that fish grow throughout life, unlike birds, where growth ceases shortly after fledging. This long period of interactions between genes and their environment in fish influences the evolution of plasticity in adult shape to a much greater extent than in birds.

A second possible research project I had considered was the evolutionary changes between domesticated plants and animals and their wild ancestors, first in the Fertile Crescent and continuing throughout the world to the present day. Here again, recent research by Leif Andersson on domesticated pigs and chickens has revealed interesting parallels with our research in Darwin's finches. As in the finches, selection and introgression have played an important role in change.[17]

17

The Magical Years as a Family on Genovesa and Daphne

> Tell me and I forget, teach me and I may remember, involve me and I learn.
>
> —ATTRIBUTED TO BENJAMIN FRANKLIN

For ten years, Nicola and Thalia accompanied Peter and me to the Galápagos as we carried out parallel research on Genovesa and Daphne. We gave them the option of staying at home with one of us, but neither of them would contemplate that idea for a moment. Although our field season overlapped their school holidays, we needed to take them out of school for some weeks each year. Their teachers helped with encouragement and a supply of appropriate materials. Fortunately, both girls easily kept up with their assignments.

On our repeated visits to the islands, Nicola and Thalia not only helped us to find nests and birds but also decided to do their own independent research; both their works were published in

FIGURE 14. Upper left: Family on Española Island, Galápagos, 1980. Upper right: With Nicola and Thalia on the *Kim*. Lower left: Family in a *panga* (small boat) with Jamie Smith (in hat), 1973. Lower right: Family on Daphne at sunset.

1979. Nicola chose mockingbirds, patiently watched them, and recorded their behavior. She discovered that the parents raised their nestlings with the help of the offspring of the previous year's brood.[1] This finding was later expanded on by Bob Curry in his PhD research with Peter. Thalia was fascinated by the doves and did a thorough study of their breeding and feeding on Genovesa as well as on Daphne.[2] Amongst other interesting aspects of their behavior, she found that they fed on the petals, nectar, and pollen of *Opuntia* flowers on Genovesa but not on

Daphne. Why not on Daphne? The *Opuntia* differs on the two islands. The plants on Daphne have long, sharp spines, whereas the ones on Genovesa have more hairlike spines. Perhaps the longer, sharper spines of *Opuntia* on Daphne were a deterrent. Thalia later studied the social dynamics of left- and right-handed fiddler crabs, feeding and breeding behavior of lava lizards, and the diet of Short-eared Owls from a collection of their pellets.

Among the books we brought, the favorites for both Nicola and Thalia were *Robinson Crusoe* and the J. R. R. Tolkien books. They called the frigatebirds that circled menacingly overhead the Nazguls, inspired by *Lord of the Rings,* and the lagoon behind the beach at Genovesa, with its dark area at the deepest part, was the home of Gollum. We have a wonderful painting done by eleven-year-old Thalia hanging in our kitchen of an imaginary man fishing on the edge of the lagoon about to catch Gollum in the form of a greedy fish. The lagoon was roughly a quarter of an acre in area and filled to a depth of about ten feet at high tide, a perfect shark-free pool safe for swimming. The children made a raft out of driftwood, modeled on Thor Heyerdahl's *Kon-Tiki,* and were regularly joined by baby sea lions and pelicans in the lagoon. Several years after we left Genovesa, a couple of storms reconfigured the beach, and sadly, this lagoon is no more.

In the early years there were few tourist boats, and we could go for as long as three weeks without seeing a boat on Genovesa. Occasionally, we would be visited by fishing boats, which would be careened on the beach of Darwin Bay on a Sunday. We were told by the fishermen that they never fished on Sundays. On one occasion, we came back from the field to find fishermen in our camp absorbed in our books. We had several, including the children's schoolbooks and novels, arranged in a temporary bookcase between two pieces of driftwood with another on top to serve as a bench. The book that engrossed them was on the history of art. During the previous year, we had introduced the children to art museums in Europe and had a book of paintings.

Never volunteering to do the washing up at home, the children almost fought to do the dishes at the edge of the ocean or the lagoon. The fragments of food would attract crowds of little fish, and soon a resident Lava Heron caught on to this and stood next to the girls, sometimes even between their legs, catching its meal of little fish. Yellow-crowned Night Herons were another source of amusement, especially their elaborate courtship displays at dusk, when they would fluff up their feathers to form an umbrella-like cover with their wings, bow down, and just as you were expecting a loud hoot, produce a ridiculously tiny peep. They seemed to be attracted to Jean Sibelius's violin concerto. When the children played it on a little cassette player in the tent at night, a heron would appear. What was it about this melancholy, brooding, far northern depiction of a Finnish landscape, so different from a hot, sultry Galápagos night, that so attracted them? Perhaps it contained echoes of their tiny peeps.

One early morning on Genovesa, we were all in the woods above the cliff netting birds, when Thalia said, "I smell bacon." Then, we all started to smell bacon. Not having had anything so exotic as bacon for months, we thought we were hallucinating. Peering over the cliff edge, we saw the Charles Darwin Research Station boat in the bay. Who was having bacon and eggs for breakfast? It turned out to be Prince Bernhard of the Netherlands, escorted by our Dutch friend Tjitte de Vries, with an entourage of several bodyguards. They left us with a present of a large, round Dutch cheese! Most boats that visited were small fishing boats or small tourist boats, with a crew of two or three, and then suddenly, one day there was a massive cruise ship in the bay, the *Lindblad Explorer*. We were invited on board for a shower and lunch. What a delight to have a freshwater shower! The captain, a kind, avuncular Swedish man, said to Nicola and Thalia, "We have absolutely everything on board. What would you like?" He expected them to say, "Ice cream." One of them replied, "Have you a G string for a violin? Ours broke." We had packed spare A strings but never thought the tough G string would break. Stumped, he apologized but promised to bring them one next month, which he did.

An amusing incident occurred several years later. Peter, Thalia, and I were alone on Genovesa. The only boats we had seen for the past several weeks had been far off on the horizon. Then a small sailing boat appeared. From the top of the cliff, we watched as a man and woman stepped ashore, she in high-heeled shoes and dressed in the latest fashion. She then posed on the beach while he took her photograph. We startled them as we appeared on the shore behind them, and in acute embarrassment, they explained that the dress and photograph were a joke. They had both studied for their medical degrees in Newcastle, England, and having completed them, sold everything, bought a boat, and were sailing back home to New Zealand. I noticed that they had come from the direction of Cocos Island. Nicola, then nineteen, had traveled down alone to join and help Tom Sherry and Tracey Werner with their research on the Cocos Finch (*Pinaroloxias inornata*), a relative of the Darwin's finches. In those days, there was only intermittent radio communication, and we had had no contact. I asked them if they had been to Cocos Island, and if so, had they seen her. To my relief, they answered yes and then said, "We thought that island was, like this one, uninhabited. Does that mean there is a 'Grant' on every uninhabited island across the Pacific!"

Throughout our long stays in the Galápagos, we never had an accident or illness. We were well equipped for medical emergencies that never arose. However, on Daphne, we had one potential mishap that could have ended in disaster. Every day we would wash in the sea by diving off the landing, swimming round to a convenient point for climbing out, covering ourselves with shampoo, and then diving back in for the final rinse. The children and I had already washed when Peter joined us, diving in. Suddenly, a six-foot-long Whitetip Reef Shark came up in attack mode from the depths, turned on its side, and opened its jaws close to Peter's shoulder. We screamed at him to come out, but fortunately, the shark abruptly turned away. Peter climbed out, and we all said it must have been the shampoo that saved him. From then on, the children called it the shark-repellent shampoo. No more diving; we switched to washing in buckets, as there were just too many sharks regularly cruising past.

Back in Montreal, we read an article in a local newspaper about a new shark repellent that divers could carry in a bag attached to their upper arm to squirt if a shark appeared. It gave the chemical formula, sodium lauryl sulfate. We rushed to the bathroom to read the ingredients on the shampoo bottle, and there it was. In high concentration, the magic ingredient seemed to work. Now, more than forty years on, we all have different versions of this story. I think it was one Whitetip Reef Shark that went toward Peter's shoulder. Thalia thinks there were several Blacktip Reef Sharks that went for his feet. Nicola thinks it was a hammerhead shark that went toward his body. We all agree on the shampoo!

This is my version of the years as a family on Galápagos, but what did the children really think? After all, they were taken away from friends and school for large amounts of time. They recorded their activities in diaries. Here are some entries Nicola made in her diary in 1975, when she was eight years old:

> *Just now I saw a school of fish. Thalia thought it was [a] Bonito but Peter [Boag] didn't think so. I pretended that a sub-marine was chasing sharks, chasing bonitos. Bonitos chasing little fish, little fish chasing tiny fish, tiny fish chasing plankton, plankton chasing smaller plankton, and small plankton looking for bacteria or rays of light, I doubt if that's true.*
>
> *Four-eyed blennies, green sea turtles, hammerhead sharks, owls, eels.*
>
> *Today Thalia and I found lots of shells and when I was looking between the rocks for shells, I found lots of bones. My precious ones are a pelican beak and skull, a boobie beak and a whale rib. Today a heron came around and Peter Boag saw two pelicans nesting in the mangroves. Thalia and I found a piece of wood and used it as a surfboard. I found a little shell which I think is a rough periwinkle but I am not sure.*
>
> *Today Thalia and I pretended that sticks were violins and we tried to play difficult pieces on them.*
>
> *On June 30 I saw tropic birds going in a screaming party. 11 birds flew in that group (that's the most I've seen). Daddy, Thalia, and I counted all the dead baby boobies and all the eggs and all the forgotten eggs in the craterlet. There were 37 dead babies, 19 eggs and 4 forgotten eggs.*

Much later, in 2014, when Nicola and Thalia were both mothers themselves, they were interviewed by Joel Achenbach for an article he was writing for the *Princeton Alumni Weekly*. He asked them what it was really like for them to live so long as a family isolated on islands and where it had led them in adult life. Thalia wrote a short, heartfelt reply: "There is always a moment in every child's life when they suddenly seem to wake up to the world, and for me it was in Galápagos at age 6. It feels like I was born there. For better or worse Galápagos has shaped my whole life, and in every direction I have taken." She became a scientist, writer, artist, and Galápagos historian. She and her husband, Greg Estes, a Galápagos resident and expert on Galápagos natural history, conducted an extensive expedition to retrace Charles Darwin's footsteps through the archipelago, using Darwin's original notes and manuscripts, some of which she transcribed for the first time, to discover exactly where he had explored. Their work on Darwin was published in *Notes and Records: The Royal Society Journal of the History of Science* (2000), in UNESCO's quarterly review (2009), on Darwin Online, and as a book, *Darwin in Galápagos: Footsteps to a New World*, published by Princeton University Press in 2009. Her illustrations, several of which adorn our walls at home, have been published in various journals and books, including Jonathan Weiner's *Beak of the Finch* and Nigel Sitwell's *Galápagos: A Guide to the Animals and Plants*. Thalia continues to conduct and publish ecological and historical studies in the islands and runs a tour agency specializing in educational natural history tours of the Galápagos. She has also written a wonderful children's book about a baby sea lion and illustrated it with her drawings. That book was never published but deserves to be.

Nicola wrote Joel a long letter (Appendix B), and I have extracted a few sentences:

I am the older of the two girls and went to the Galapagos every year with my parents and sister from the ages of 8 to 18. Quite simply, it was magical. I don't know if you have been to the

Galapagos Islands, but for me they are like what the Celts call "thin places"—places where the veil between heaven and earth is frayed. My sister and I were very lucky to be able to spend a few months each year there. . . . I don't remember ever being bored. We spent our days exploring whatever island we were on, swimming, inventing games, reading, and the older we got, the more we helped our parents with their research work. During the school year that we spent on the islands, my mother would teach us during the hottest part of the day, and the rest of the time we had to ourselves.

Nicola studied English and comparative literature as an undergraduate. She spent a year abroad in Nepal, became fluent in the language, and developed an interest in oral literature spoken in the remote villages. She hiked with a female companion to the Everest base camp. She fell in love with the classical sitar and returned to India after completing her undergraduate degree to study the instrument (chapter 19). She excelled in this, giving concerts in Europe for a year before deciding that she would like to study medicine. Now she is a family doctor with her grandfather's skill at diagnosis.

After our second summer visit to the Galápagos, in 1975, when Nicola and Thalia were ten and eight years old, respectively, we decided to visit Machu Picchu in Peru on our way back to Montreal. Flying directly from sea level to Cuzco at 11,151 feet elevation, we all felt the altitude and had to rest. Traveling by train the next day to Machu Picchu, a mere 7,972 feet above sea level, gave us welcome relief from the symptoms of tiredness and headaches. In those days there were few tourists, and we were the only guests in a tiny, four-room hotel. Staying in this hotel allowed us to walk onto the site alone the next morning and watch the sun rise above this ancient Inca citadel.

The ruins sit high on a plateau surrounded by terraced slopes; a steep pinnacle towered above. The children were off, scrambling upward on a tiny path to get to the top, with us following close behind. From there, perched above the ruins, we had a bird's-eye view of the layout and magnificent views of the surrounding mountains. We still had the whole area to ourselves to explore. Buildings of massive stone walls were so sturdy that generations of earthquakes had not disturbed them. These gigantic stones are set together so closely that not even a thin knife could be inserted between them. The most impressive area was the remains of a temple dedicated to the sun near the Intihuatana stone. *Intihuatana* in the Quechuan languages means "place to which the sun is chained." As with so many structures we had visited in England and France (chapter 13), the stone was another astronomical observatory but this time with a difference. Machu Picchu is located at 13.2 degrees south latitude, and the Intihuatana stone is set at a slight angle of exactly thirteen degrees; thus the sun is directly above the stone at the equinoxes (around March 21 and September 23). The arrangement of the stone, with its inclination of thirteen degrees, suggests that the equinoxes might have been important dates for planting and harvesting agricultural products.

We had read that although the Inca made children's toys with wheels, they never used wheels on structures for transport; instead they relied on llamas, which were much more efficient in this mountainous country with its narrow trails. We walked some way along one of these, the Inca trail that connected Cuzco to Machu Picchu, watching butterflies and hummingbirds feeding on the alpine flowers, and were lucky to come across a coatimundi. At one point, we had thought we would walk and camp along the entire four-day route from Cuzco to Machu Picchu, but the presence of the Shining Path (or Sendero

Luminoso) terrorist group in the area at this time made it unwise. At lunchtime the train arrived, bringing a flood of tourists to visit the site for a few hours, and we returned with them to Cuzco at the end of an unforgettable day.

The following day, we explored Cuzco, a three-thousand-year-old city, our altitude sickness now cured. The highlight for Peter and me was a medical museum with a series of skulls on display. They had been trepanned a thousand years ago by Inca surgeons in the identical manner that had been found in seven-thousand-year-old Neolithic burial sites in France, another sign of similar practices that had evolved independently in different areas of the world. We visited the cathedral in Cuzco's main plaza. It possessed a 1753 painting of the Last Supper by Marcos Zapata inspired by Leonardo da Vinci's well-known painting. The Cuzco painting is famous because it shows Jesus and the apostles about to dine on a guinea pig! Walking through the markets and admiring the local alpaca serapes and blankets, we were hassled by men offering to buy Thalia's fair hair. One bargained with Peter: a poncho for her hair! On the way back, we were invited into a small house with guinea pigs running around the dirt floor along with hens, both food items for dinners.

18

Teaching and Research in Princeton

Education is the kindling of a flame, not the filling of a vessel.

—SOCRATES

We had settled into life in Ann Arbor at the University of Michigan. We had exceptional group discussions with graduate students, postdoctoral students, visitors, and professors, and the opportunity to do research on Genovesa and Daphne. The children were in an excellent school and had good violin teachers. The university music school was another big attraction, and we relished the almost weekly concerts from symphony orchestras around the world with such artists as Nathan Milstein and Murray Perahia. Our house had a stream at the bottom of the garden and woodland beyond with small ponds containing fairy shrimps, to Thalia's delight. They so enthralled her, she would later study them in Canada and Galápagos. Then, in 1985, came a phone call to Peter from Robert May, asking him whether he would be interested in applying for a position at Princeton. This was attractive to Peter, and there would be a position for me as a research scholar, with the opportunity to do research and teach. Robert May, John Terborgh, Henry Horn, Dan Rubenstein, and John Bonner, who I had first met as an undergraduate at Edinburgh, would be our colleagues. Importantly, the invitation had come at an appropriate time for us as a family. Thalia was about to leave for the University of California in Santa Cruz, Nicola was in her second year at Dartmouth, and I now had my PhD. Another big perk was that Peter and I could do double duty teaching in the fall term, freeing us for research in the Galápagos in the spring when the finches breed. This time it did not take us long to decide to accept the offer. For the next many years, we happily combined teaching with research at Princeton University. Even after we officially retired and gave up teaching, the university has kindly allowed us to keep our office. We continued field research for many years and today are collaborating with geneticists using the now thousands of blood samples on individual marked and measured birds we had collected over more than forty years. The results, finding genes responsible for the changes we observed over the selection events, are incredibly exciting, and Peter and I often wish we could have two or three lifetimes to explore all our questions.

My first impression of Princeton, New Jersey, was that it is a village like Arnside. When we first arrived, the small, European-style shops carried essential goods and wares. A hardware store that on first sight was chaotic had everything; there were two shoe shops, clothes shops, and a small pharmacy. These were gradually replaced by large chain stores, and now Nassau Street, the main throughfare, is a row of takeout food stores. The delightful fish shop is still there, tiny, a narrow passageway to the counter, with excellent high-quality fish. When we first arrived, the elderly man who owned it loved classical music, and we bought our fish to the strains of a Beethoven overture or a Mozart concerto. The store today is identical in all ways but without the music. There were pockets of racism in Princeton; once, our daughter and her dark-skinned husband were turned away from a restaurant with the feeble excuse that he did not have a tie—nor had many of the other diners! In contrast, the toy store had a window full of children's books with the theme *We are all immigrants*! On our walks along the former canal towpath, in the woods, and by the lake, people pass with a cheery greeting. As in Arnside, it appears that 50 percent of them own dogs. Discussions with the person checking out groceries at our local organic food market turn to the latest opera in New York, a chamber music concert in Princeton, sightings of a rare white deer, the first bird migrant or wildflower of spring, and so forth, with people patiently waiting in line behind us or joining enthusiastically in the conversation.

Peter and I fully enjoyed lively interactions with our colleagues, students, and postdocs, as well as the numerous seminars, talks, concerts, and discussions, on all topics from biology to literature, politics, and music, that the university has to offer. We found a delightful small house, within walking distance of the university and trails, so we rarely use the car. Over the years, we added solar panels and heat pumps that keep us comfortable with minimum use of fossil fuels. When we first arrived, Killdeer nested in the field across the road, and American Kestrel and Wood Thrush were commonly seen, along with numerous butterflies and moths, including the magnificent Luna Moth. Sadly, all are gone from the vicinity.

Nicola and Thalia, both away at university when we arrived, returned during vacations, often accompanied by friends. Then, as they grew older, they were accompanied by husbands and children. We had kept Nicola's and Thalia's children's toys and books in our finished basement, a delight for our grandchildren, who felt it very special to be playing with their mothers' old toys and reading their books. All four grandchildren are now fully grown; they still visit us, and our grandson particularly enjoys driving our now thirty-seven-year-old Volvo.

The opportunity to do double teaching in the fall and spend three to four months each year in the Galápagos during the birds' breeding season benefited our work enormously. I have always enjoyed teaching and interacting with young people, with their lively flexible minds and energetic enthusiasm for exploration. Princeton gave me this opportunity. I taught a third-year animal behavior course with a strong bent toward evolution. Peter and I set up a course for first-year graduate students and named it the Journal Club. We divided recent scientific journals among the incoming graduates and ourselves to cover the specific interests of the graduate students. This usually amounted to about seven journals each. The goal was to survey the papers and bring to everyone's notice articles of relevance and then to read and discuss one or two articles in depth. This worked superbly well, with the additional advantage that the incoming cohort of graduates found out about each other's research plans and interests.

Princeton has two requirements for undergraduates; one is a freshman seminar, and the other an independent research project with a thesis. For the latter, all professors in our department meet with the class one evening early in the third year to describe the research in their laboratory. I had planned a little three-minute speech that was to follow several others, but seeing a group of warm, sleepy, post-dinner faces in front of me, I mentally tore it up and instead said, if anyone has a passionate interest in anything, I will be willing to help you and will be the first to tell you if your interest is completely out of my area. Suddenly, several people sat up, eyes sparkling, and from then on, for the rest of my teaching career, I had many wonderful undergraduate students to interact with while they were doing their independent research on a topic of their choosing. Their topics ranged widely: the origin of flight, the transition from hunter-gatherers to pastoral agriculturalists in communities of Indigenous South American people, communication among whales in the Saint Lawrence Seaway, migration of turtles, the effect of prematurity on development in babies, and many others. Each week, I looked forward to my hourly one-on-one meetings with young people who were following their passion.

I labeled my first freshman seminar course "The Origin of the Human Condition." Later this became the name of a course for nonscience majors that I taught with Shirley Tilghman, who was then a prominent molecular biologist and was soon to become president of the university. Shirley was a better and more experienced teacher than I was, and I learned much from her over the three years that we co-taught the course. I recently found my handwritten introduction and outline for my part of the freshman seminar. I used two required texts, *Human Diversity*, by Richard Lewontin (1995), and *Animal Minds*, by Donald Griffin (1992), supplemented by many additional readings of research papers by people such as Eric Kandel, Luigi Luca Cavalli-Sforza, and others.

My introductory outline built toward a set of questions:

> *One of the outstanding findings in the last hundred years is that all forms of life from bacteria to humans have a similar mechanism*

of heredity. Furthermore, we have in our bodies some genes that give rise to enzymes that serve the same function as those found in bacteria, plants and animals. This fact suggests a common origin but does not explain the infinite variety of life around us. Why are we different? How did we come to exist? What does it mean to be human?

The course fell naturally into three parts. First, we considered the source of diversity—that is, the genetic mechanism of heredity. We discussed the interaction of genes with the environment and how and why genetic combinations change over time. Then we got into how new species are created and why they go extinct. Second, we deliberated on how *Homo sapiens*, a species capable of language and conscious thought, came to be. Then we explored why, in the past ten thousand years, humans moved across the globe. Third, we discussed the consequences of being human, with our capacity to conceptualize ideas and communicate them to others. Questions that arose in rudimentary form when I was a child were now honed and explored in adulthood.

In the course I taught with Shirley, we added a laboratory and required the students to do an independent project entirely conceived by themselves. I will describe one of the many projects to illustrate what can be achieved by allowing young minds to explore their own ideas. One group of three decided to test the concept of natural selection leading to evolution. They asked the researchers in the molecular yeast laboratory what they used to swab down their benches to avoid yeast contamination. They were told, a dilute solution of Clorox. They diluted Clorox at different concentrations and found that the wild yeast they had collected off trees on campus were killed at very low concentrations, whereas the lab yeast survived even higher concentrations than was used to swab the benches! The lab technicians had been innocent agents of selection, unknowingly selecting increasingly resistant strains of yeast.

At the time, there was much discussion of the idea that body odor preferences determined mate attraction in humans as well as in mice.[1] The proposed mechanism was a link between odor receptors and MHC (major histocompatibility complex) molecules that are involved in immune responses. Disassortative mating—like breeding with unlike—would lead to a maintenance of diversity in natural populations. This idea appealed to many but was contradicted by others who could find no such relationship. The young students got excited about these ideas. Could smell be a possible factor in their choice of a partner for life? They were so keen to test this that for the next three years, in the last two weeks of class, we conducted what became known as the T-shirt experiment. Shirley and I washed new T-shirts in bland soap, one for each student, told the students not to wash that morning but to run several times round the racetrack wearing their shirt until they got sweaty, and then we conducted a double-blind smell test to see which T-shirts were preferred. To their disappointment, the results each year showed no preferences. Yet, the whole experiment not only taught the students how to perform PCR (polymerase chain reaction) analysis of their saliva to determine their own MHC genetic signatures but also led to many discussions ranging from environmental factors that could alter the results to how plausibly exciting theories might blind a person to the interpretation of results. The students cheekily asked me to do the experiment on Peter and myself. I did not like Peter's sweaty smell any more than he liked mine, but we did discover we differed at the MHC genetic locus.

In Montreal, we had often hosted open-house parties for everyone who attended the weekly seminar in the Department of Zoology. This we continued to do in Ann Arbor and now in Princeton, where hosting is rotated among those of us living close to the university and having it in the university. Now we are retired, these occasions become particularly enjoyable, as we have fewer opportunities to meet the graduate students and postdoctoral fellows.

19

Interlude in Nepal

> Learn from yesterday, live for today, hope for tomorrow. The important thing is not to stop questioning.
>
> —ALBERT EINSTEIN

Before we moved from Ann Arbor to Princeton in 1986, we requested and were given a year's leave of absence. This enabled us to spend more time in the Galápagos but also gave us opportunity to explore new places. We had the perfect excuse for unusual travel: a daughter in Nepal. Nicola was spending a study-abroad year in Kathmandu, and that was the magnet that drew us to visit Nepal for the first time. We were excited to visit Asia, not only to see firsthand the very different biodiversity in a new environment but also to gain insights into a culture different from our own backgrounds. It was a forerunner of experiences to come in retirement.

Nicola had spent the summer learning Nepali in Madison, Wisconsin, and by the time she departed in September had acquired an excellent grasp of both the spoken and written language. Her independent study goal was to record and transcribe the oral literature in a remote region of Nepal where a hydroelectric dam was to be built and the local people were to be relocated to Kathmandu. She hoped her efforts would give them access to their own literature after their displacement. Knowing the strong similarity between oral literature and music, she took her violin with her. While there, she fell in love with the classical sitar. This led to sitar lessons from Uma Roy, a talented sitar artist, and eventually to playing in concerts in Europe and the United Kingdom with Uma and the renowned tabla accompanist Hom Nath Upadhyaya.

I can well understand her love of this music, having heard on the radio a mesmerizing piece with extemporizations played by cello and sitar. The sitar has a resonating gourd and long neck with twenty strings, thirteen of which are sympathetic strings situated below the others. Often tuned to the note D, its sound has a hollow echo with reverberating harmonic overtones. A raga, a typical conversation between instruments, starts simply, intensifies in emotion, and has long stretches in which the player improvises, with the result that no two compositions are ever the same. The instrument may have originated from the similar Persian *sehtar*, or there may have been a fusion and further refinement of both. Scott McLennan, in 2016, wrote a preview of a concert by Zakir Hussain, a tabla player from India, and Japanese drummer Seiichi Tanaka, to be held in the Sanders Theatre in Cambridge, Massachusetts. In it, he quoted Hussain giving a moving example of the harmonious creative improvisation between cultures: "Music is universal, it all begins with the pulse. Japanese drumming is similar to Indian drumming in terms of ancient traditions. When you approach these ancient traditions with mutual respect, doors open, and you let in the music."[1] I think of this fusion as illustrating the inventiveness that becomes possible through collaboration, cooperation, and respect on all topics, art, science, and politics—a theme that was to stay with me throughout all our travels.

Nicola accompanied us to Chitwan National Park in the Terai, an area of tropical rain forest dominated by Sal trees. The visit was a chance to see a variety of mammals, birds, and reptiles, the most famous being the endangered one-horned Indian Rhinoceros and the Bengal Tiger. From the back of an elephant, we spotted the elephant grass waving as, we were told, a tiger stealthily walked through it. If true, it was the nearest we got to a tiger, but we did see several rhinos. We missed seeing a Sloth Bear, Honey Badger, or otters but had magnificent views of wild junglefowl that reminded me of my mother's hens, and close-ups of the Himalayan Monal (related to peafowl), a paradise flycatcher, and the Bar-headed Goose. The goose was particularly interesting to us because we knew of the genetic modifications that produced its larger lungs and a type of hemoglobin that can bind more oxygen, allowing it to migrate twice a year over the Himalayas at an altitude of 23,600 feet.[2]

Another interesting genetic adaptation occurs in the Tharu people in Chitwan: a high proportion of the population has the alpha-thalassemia gene for protection against malaria infection, a disease rampant in this region.[3] We watched a stick dance in which Tharu men move slowly in a circle to the beat of several drums, watched by an outer ring of young women chaperoned by their mothers. At every third beat, the men turn back in unison to hit their neighbor's stick. If one misses, he drops out of the ring. The contest continues until one man remains, the winner, who can then choose one of the surrounding young women. The ritual would surely have selected for the strongest malaria-free male. Perhaps this contributed to the reason the variant has become so common in the population; they have lived in the forest for centuries.

We hiked in the Himalayas while Nicola stayed in Kathmandu. In the previous months, she had hiked as far as the Everest base camp with another woman, both wearing Tibetan costume. Her boots had been stolen on the bus between Kathmandu and the trailhead, forcing her to walk for a month in running shoes. She and her friend were both fluent in Nepali, and the owners of the teahouses they stayed in along the way thought they were from Nepal and spoke with a Kathmandu accent. Each night, they were well looked after and given strict instructions about who to stay with in the next village. When the time came to make our arrangements, our roles were reversed, and for the first time, our daughter was concerned about her parents. Nicola gave us strict instructions and told us not to even consider hiking alone. "Mum, you are not even to think of wearing jeans or pants, instead a long woolen skirt that sheds the rain. A skirt acts as a tent when washing in streams and is considered polite among owners of teahouses."

We had to have a guide, she told us, and knew the perfect man, Pasang Sherpa. We should go into the center of Kathmandu to a trekking company in a certain building, and we would find him there. This we did. The first person we met was Pasang, and we arranged to trek with him for the next three weeks. Returning triumphant to Nicola and feeling pleased with ourselves, we found out he was the wrong Pasang. Apparently, every third person is Pasang Sherpa! However, our Pasang was highly experienced and overqualified for us. He had been Tenzing Norgay's right-hand man, had climbed Everest three times without oxygen, and even more helpful for us, he had worked for several years as an interpreter for anthropologists in the Langtang region. He knew all the teahouses and their owners, villages, and trails in the area. Later, we discovered he knew all the birds, plants, and animals, and he helped us to identify them from our books. He was a fount of knowledge about the medical properties of plants. Peter was prone to nosebleeds, and each time he had one, Pasang found the appropriate plant nearby to staunch the flow of blood. It was interesting to learn that Pasang had climbed Everest without extra oxygen, because he may have benefited from a variant of a gene that was shared with our ancient ancestors the Denisovans.[4] The gene is called *EPAS1* and affords those who have it the ability to cope with low oxygen levels at high altitude. As more than 87 percent of Tibetans and Sherpas carry this genetic variant, it is highly likely that Pasang also benefited from it. We never knew Pasang's age, but he said he learned English while being a Gurkha in the army.

After buying footwear for Pasang and a few other needed things, we set off by bus from Kathmandu to a trailhead. Then, walking first through forests of oak, maple, and pine, interspersed with villages surrounded by terraces for growing rice and other crops, we began our three-week trek. Blooming rhododendrons and *Daphne* bushes became common higher up. After several days, we were in high alpine meadows and finally in snowfields and permanent ice high on the slopes of the 23,734-foot-tall Langtang mountain. We took it in turns to carry two packs among the three of us, stocked only with essentials and three sleeping bags. We stayed in teahouses along the way, mostly wooden or stone huts with a central open fire, the smoke curling up through a hole in the thatched roof, and we slept three in a row in our sleeping bags on a wooden shelf. We washed in cold streams along the way. Pasang dictated where we should wash and where we could and could not fill our water bottles. At one point, we reached a ridge overlooking a deep, steep, narrow gorge. On the opposite side was Tibet, so close that we could see men and women working in a field and hear them talking.

FIGURE 15. High up in the Langtang Valley, Nepal, with Peter and Pasang Sherpa, 1986.

Halfway up we reached Kyanjin Gompa, at 12,700 feet. That was the one place where we exchanged sleeping in teahouses for a simple hotel built of rhododendron wood. Two young Swiss, Regula and Stefan, joined us on our trek. Regula pounced on the chocolate bars left by a mountaineering expedition and now for sale, bought a pile of them, and gracefully shared her loot with us over the next few days. Near this tiny collection of houses were a cheese factory and a monastery. The cheese factory had been built by the Swiss for the local yak herders. Complete with Swiss wooden vats and heavy wooden cheese presses, they produced delicious wheels of Swiss-style hard cheese made from yak milk, to be carried down the mountain on the backs of porters and sold as far away as Kathmandu.

On the monastery balcony was a row of *kangling*, long trumpets or horns made from human tibias. Pasang looked at Peter's long legs and said, "Be very careful, those legs would make the most beautiful *kangling*." On our way toward the monastery and at intervals along the trail, we passed stupas, burial mounds faced with stone, which Pasang guided us around clockwise to pay our respects.

From Kyanjin Gompa we climbed higher, seeing choughs, ravens, eagles, and Lammergeiers (or Bearded Vultures), and on the scree slopes, pikas. Alpine Accentors were feeding on insects and other arthropods on the snow, looking and behaving very much like Rocky Mountain rosy finches. Like them, the accentors migrate vertically up and down the mountain, being physiologically adapted to high altitudes. It was on this segment of our journey that we saw a Red Panda being chased down a slope by angry ravens. Then, after a long day wearing goggles to trek across snow and ice in brilliant sun, we reached our destination, a tiny stone hut. That night the snow whistled and blew through the gaps between the stones. An Alpine Accentor flew in and spent the night with us in relative warmth, while Peter and I could keep scarcely warm in our down bags. Pasang's bag was thinner than ours, so we covered him with a space blanket. The result was that he got so hot and perspired so much that he woke the next morning soaking wet. "These modern inventions, uh," he said in disgust. We retreated down the mountain so he could dry his bag in the sun, not sorry to spend that day resting in a beautiful alpine meadow. Then we made our long descent back to Kathmandu, hoping to escape the onslaught of the monsoon season. The first rains hit while we were on the bus back to Kathmandu. We were stopped by not one but six landslides and spent hours helping to clear rocks to allow the bus to crawl its way slowly forward.

During those three weeks, we were alone most days. An exception was our meeting with a couple of Englishmen in a teahouse; as I described in chapter 1, my mother had rescued one of them in Arnside when he was a boy. Pasang knew everyone we stayed with. He told us that every man from these tiny communities had to walk ten days to obtain a wife to avoid inbreeding. To me this was striking, because we had just spent time in Switzerland where, in the mountains, there were villages of highly inbred people showing some of the typical detrimental complications. Why did these two isolated mountain communities differ in such a fundamental respect? It seems that prehistoric small, agricultural family groups throughout the world were aware of the problems of inbreeding, because genetic analyses show evidence of large mating social networks, like the one in Nepal. Even in England's Lake District, there is a notice at the entrance of a tiny stone shepherd's church that proclaims: "NO FIRST OR SECOND cousin marriages allowed!" The church (Saint Anthony's, built in 1504) is situated in the once remote area of Cartmel Fell. Why, I wondered, is Switzerland the exception? The desire to keep land and wealth in the family was the explanation I was given. Pasang was proud of his Tibetan ancestry and polyandrous culture. He explained that in his case, two brothers had walked for ten days to find and marry his mother. He also swore he knew which one was his father. Pasang's explanation of polyandry was that a Sherpa leads a dangerous life, and a brother acts as a replacement father in the event of the other one dying.

Pasang, Regula, and Stefan remained our good friends. Letters and cards passed between Pasang and us at Christmas until the year he died.

This visit to Nepal and its introduction to genetic, morphological, and cultural differences in isolated mountainous habitats whetted our appetite for more visits to Asia. They were realized in retirement.

PART V

Retirement and Research

20

Retirement

> Learning is the only thing the mind never exhausts, never fears, and never regrets.
>
> —LEONARDO DA VINCI

Some years are much more eventful and momentous than others. For us, the year 2005 was pivotal, both domestically and professionally, and it began with the discovery that Peter had colon cancer. This resulted in an immediate operation that eliminated the cancer but unfortunately left him with half a colon, which he now refers to ruefully as his "semi-colon." In the meantime, Daphne Island was experiencing a prolonged drought, and we were eager to return and find out which birds had survived and which ones had died. Our good friends Uli Reyer, Ken Petren, and Lukas Keller kindly substituted for us and did a superb job of finding the survivors. While they were there, the rains finally returned after two and a half years of drought and produced a torrential flood. Thalia then followed with a short visit. She is an excellent observer and managed to document and read the bands of almost all the survivors and record the songs. These valuable observations were crucially important for the character displacement result discussed in chapter 16.

Dolph Schluter and Trevor Price thought we were retiring in that year (or should retire!) and organized a magnificent symposium, a "Grantfest," at the University of British Columbia. Here we renewed acquaintances with ex-graduate students and friends, had lively and fun discussions, and heard about their most recent research.

Peter had recovered enough by the summer for us to join the World Summit on Evolution on San Cristóbal Island, at the Galápagos campus of the Universidad San Francisco de Quito (USFQ). This was organized by Santiago Gangotena, Carlos Montúfar, and Carlos Valle. It was an unusually interactive meeting with time to have long enjoyable discussions with participants from around the world. At the conference, Peter and I were made honorary citizens of San Cristóbal (*huéspedes ilustres*), as was Charles Darwin (posthumously), and at the end of the meeting, we were awarded honorary degrees from USFQ, together with Lynn Margulis, for her outstanding research on symbiosis, and Thomas Kunz, an expert on Ecuadorian bats. Sadly, both Lynn and Thomas have since died, Thomas in 2022 from Covid-19.

In this year of ups and downs, we heard the sad news that our good friend Jamie Smith had died of cancer. We had known him since he was twenty-three. He was a frequent visitor to our home and a wonderful companion, especially in 1973, during our first year of Galápagos research (chapter 15). We always exchanged letters, e-mails, and books, and that year we visited Vancouver and saw him and his wife, Judy, shortly before he died. He had just managed to complete his book on the conservation of small populations, which was published with his uniquely expressive drawings.[1]

There is no mandatory retirement age, and some college professors continue "in harness" until they drop. We knew we would retire from our positions at Princeton when it became apparent Peter was having increasing difficulty hearing. The only question was when. We enjoyed teaching and were both reluctant to give it up. Yet, we also knew jobs for young scientists were scarce, and our retirement would release two spaces, and we were very much aware that a vibrant science department needs energetic young people with fresh ideas. We took the decision to retire in 2008, at age seventy-two. The department in the meantime has hired several young, brilliant, energetic professors—more than half of them women. Our official retirement was well worth this outcome, and we are still in the department very much enjoying our colleagues' invigorating company.

Retirement meant retirement from teaching and committee work, not retirement from research. In fact, we continued intensive fieldwork until 2012, and in 2023 we are still collaborating with geneticists using the tiny blood samples we collected from each finch over the years and analyzing data. We are also involved in seminars, international graduate teaching workshops, and other enjoyable research and teaching commitments around the world. Princeton University has kindly allowed us to keep our offices in the department, and we have access to the library and help with computer technology. And we have the young, lively, enthusiastic professors as companions to keep us up to date and on our toes.

On one of our return visits to the Galápagos, Peter and I went alone to Genovesa to do some song-playback experiments and collect blood samples for DNA analyses. The small boat we had hired from the Charles Darwin Research Station had engine trouble. We were delayed in arriving and landed just before sunset, then rushed to get all equipment, food, and water on land before it became too dark to safely climb the cliff to set up our tent. For that night, we put our small dome tent at the back of the beach and went to sleep. A few hours later, wump, wump, wump, an enormous male sea lion, mad as hell, appeared at the tent entrance. We escaped out of the other side. This was no Galápagos Sea Lion, which usually departs if you stand your ground; he was huge and incensed, possibly because we were in his territory, and he started to chase us. We knew he could outrun us if we ran straight but would be much slower if we ran in circles, so round and round the tent we went, two naked figures holding hands, silhouetted against a full moon, with a gigantic, enraged sea lion after us. Eventually, exhausted, he flopped off and climbed on top of our supplies, leaving a foul-smelling mess on the tarpaulin for us to clean the next morning. Peter was back in the tent and asleep in five minutes, but there was no more sleep for me that night. It turned out this gigantic beast was a rare rogue South American Sea Lion that had temporarily entered Galápagos waters. I imagined him swimming up the cold Humboldt Current from Argentina. South American Sea Lions are among the largest, with males measuring up to nine feet two inches in length and weighing as much as 770 pounds. Our sea lion certainly looked this size.

Some of our visits to Galápagos since 2012 have been greatly aided by Carlos Valle. Carlos was born on Galápagos. He did his PhD study with us in Princeton on the unique Galápagos Flightless Cormorant and is now professor and dean at USFQ. Apart from his skill in administration, Carlos is an excellent and observant field researcher and has stepped in several times to help on Daphne, both before and after our official retirement. In 2018 he was invaluable when, with Leif Andersson's group, we ambitiously took the elaborate Oxford Nanopore equipment to Galápagos for the first step in genome analyses conducted in the laboratory on San Cristóbal.[2]

Full retirement from teaching at Princeton was preceded by a year's furlough leave in 2008. We decided to spend it in Vancouver at UBC. We drove across Canada, visiting our old haunts on the way, in our then twenty-one-year-old Volvo (now, in 2023, it is thirty-seven and still running beautifully). These stops included camping and hiking along the north shore of Lake Superior, walking trails in the Rocky Mountains, and making a quick visit to the Ashnola area, where we became engaged in 1961. Arriving in Vancouver, we settled into Saint John's College, which provided a residence for graduate students, postdoctoral fellows, and visiting scholars. The only stipulation was that we had to dine and converse with the residents—not a chore but extremely pleasurable. The master of the college was Timothy Brook, a professor in the History Department who specializes in Chinese history. While we were at the college, he finished his delightful book *Vermeer's Hat,* in which he discusses the influence of global trade in the seventeenth century through objects in Johannes Vermeer's paintings.[3] I have bought many copies of his wonderful book to give as presents. The college was within easy walking distance of the Department of Zoology, and we divided our time between science in the daytime and art, history, and music in the evening.

The first year of our true retirement was 2009. It coincided with the second centenary of Darwin's birth and the 150th anniversary of the publication of *Origin of Species.* Because of our work on Darwin's finches, we were inundated with requests to speak at symposia. Ninety-five invitations in total arrived, of which we managed to accept forty-five, all squeezed in before and after some weeks of intensive fieldwork on Daphne.

The Darwin year celebrations of 2009 began in September 2008 with a symposium held in Bogotá, Colombia, where we stayed with Margarita Ramos, a former Princeton PhD student, and her mother. As we walked up the hill to give our lecture at the Universidad de Los Andes, we were reminded of the strong literary tradition in Colombia when we passed a group of students reading excerpts from their poems to an audience in a poets' corner. The custom was the fusion of different cultures, we were told, particularly Spanish and Indigenous Colombian, and that this interest was now encouraged in schools, universities, concert halls, and art galleries.

Following our seminars, we were fortunate to visit a nature reserve in the *páramo,* an area of land donated by Hendrik Hoeck, a former director of the Charles Darwin Research Station, an expert on the biology of African hyraxes and a very good friend. Just before leaving the tree line to enter the more sparsely vegetated area of the *páramo* tundra, replete with large rosette plants, shrubs, and grasses, we watched wild guinea pigs and a Sword-billed Hummingbird. This spectacular bird has a four-inch beak on a five-and-a-half-inch body, which it was using to probe a trumpet-shaped flower, somewhat reminiscent of cactus finches in Galápagos. Peter and I immediately wondered what genes were involved; were they the same as those seen in the development of finch beaks? With global warming, the *páramo* ecosystem and its inhabitants, both plants and animals, are rapidly shifting location and evolving to adapt to a different ecology.

We ended our visit to Colombia at the spectacular gold museum with its collection of pre-Columbian gold, pottery, and textiles. The history of Colombian Indigenous peoples mirrors what happened elsewhere throughout the Americas. When the

Spaniards arrived, in 1509, they found 1.5 to 2 million people living in the area that is now Colombia. Many of these people were agriculturists and metalworkers, whose advanced skill in mining and metalcraft could be seen in the objects displayed in the museum. Interestingly, they had communal land laws very like the administration of local common land in England's Lake District. As happened elsewhere, Europeans imposed a reservation system, allowing Indigenous peoples access to only a restricted area of land. The Roman Catholic missions were granted jurisdiction over the lowland peoples in 1887, and over the years, evangelization and education by the church extended elsewhere. As pressure for land increased, white settlers encroached on the reservations largely unimpeded. During the past twenty years, the plight of Indigenous peoples worldwide is becoming widely known but is still far from being resolved.

The year 2008 was only the beginning of our trips around the world in retirement, which took us to all continents except Antarctica. We took advantage of invitations to university lectures, conferences, symposia, teaching international graduate courses, and visiting professorships to explore new environments and learn about the local culture and history by visiting museums and art galleries, going to concerts, and hiking in the different ecosystems.

The question at the back of my mind as we traveled across the globe was: Why, with our ability to communicate across cultures and time so well, are we still faced with racism, xenophobia, extreme inequalities, and war? How do different cultures confront these problems, and what can we learn that might throw light on mitigating such huge global challenges as extreme inequalities among peoples both within and between countries, climate change, deterioration of the environment, and loss of biodiversity? As I look back at our travels, I find that my questions and answers changed with the continents. Following this theme, I will arrange our visits by continent rather than mention them in chronological order, but first I will describe our experience living with an Indigenous culture in the Amazon and our three-month visit to northern Australia, which shone a light on problems seen throughout the world.

21

Indigenous Peoples

> Only when the last tree is cut, only when the last river is polluted, only when the last fish is caught, will they realize that you can't eat money.
>
> —NATIVE AMERICAN PROVERB

We live in a world largely disconnected from the Indigenous peoples who once dwelt on the land we now occupy. Growing up in Arnside, I was fascinated by the history of the Lake District, our ancestors, and how they managed through cooperation to live for thousands of years in relative harmony with minimum impact on their environment (chapter 6). Human cultures differ in their environments, cultural analogues to the biological evolution that I studied. There are examples of parallel cultural change, as there are examples of parallel biological evolution. Examples of change in parallel or even convergence suggest common causes and a possible key to understanding them. With this as background, I was keen to take any opportunity that came our way to see Indigenous peoples at firsthand in the environments that had made them what they are. The first opportunity came in Ecuador.

Groups of Indigenous peoples still live in the Amazon in much the same way as they have for thousands of years. In 1991, Peter, Thalia, and I were fortunate to live for a few days with the Siona people, a hunter-gather group in the Cuyabeno reserve in Ecuadorian Amazonia. We were invited by our friend Tjitte de Vries, a professor in ecology at the Pontificia Universidad Católica del Ecuador in Quito, who was active in field research in the Andes, Amazon, and Galápagos. In 1989 to 1990, he had studied a subspecies of the Black-mantled Tamarin (*Saguinus nigricollis graellsi*) in Cuyabeno and while there had befriended the head of the Siona, Victoriano, as he now calls himself. (He did not tell us his Siona name.) Victoriano's first encounter with a non-Indigenous Ecuadorian, he said, was in 1981, when he was sixty years old, only ten years before we met.

Tjitte, accompanied by his wife, Cecilia, and their three children, drove us from Quito to the headwaters of a river where Victoriano and his grandson met us in a dugout canoe made from a complete tree trunk with a small modern outboard motor attached. All eight of us piled in, sitting in single file, Victoriano standing in the prow, and his grandson steering the outboard motor. As we glided slowly down the river, the forest trees on either side were perfectly reflected in the still water, giving us the sensation of traveling upside down. Bright blue *Morpho* butterflies flitted across, disappearing instantly against the dark background as they folded their wings. Eventually, the river opened out into a network of lakes, the forest in this area being seasonally flooded, with islands of *terra firma* here and there. The half-submerged trees were smothered in epiphytes, and through the leaves we caught glimpses of many birds including Blue-and-Yellow Macaws and a group of Hoatzins perched on branches overhanging the lake, looking like prehistoric creatures. Young Hoatzins have claws on their wings, enabling them to

FIGURE 16. Upper: Tiputini River, Ecuador. Middle: Victoriano and his grandchildren. Lower: Siona peoples' houses.

crawl up into the vegetation. Their food, entirely vegetarian, is fermented by bacteria in their crops, just as the plants break down in the rumen of cattle, and like cattle they emit a foul, fetid smell. Later we were to find river dolphins, an otter (*Lutra longicaudis*), and several species of primates, including howler monkeys.

On a patch of raised dry land, Victoriano showed us around one of the Siona's communal houses, a large wooden structure built on stilts and with a thatched roof, which accommodated several families. He told us they occupied the house for a few years, until it became inundated with rodents and insects. They then moved on, built another, but returned at intervals to harvest the small garden of medicinal plants, yucca, and papaya. Only six years before we met Victoriano, he had guided a botanist, who taught him the Latin names of the plants and animals, which he used in preference to Spanish or English names when showing us around. He pointed out *Atta colombica* (leaf-cutting ants), a clear favorite of his. We watched, engrossed, as a line of many thousand ants wound down the tallest tree, each carrying a piece of leaf twice its body size, to eventually disappear down a hole in the ground that led to their fungus garden. Victoriano explained how his people harvested their food, which was mainly fruit and fish, with the occasional meat from a tapir or monkey. He said he was seventy-one years old, but to us he looked no more than forty-five. His wife, of the same age, was expert at spearing fish and caught a huge arapaima—dinner for all of us. My impression was that the transition from hunting and gathering to an agricultural way of life was more a continuum than an abrupt switch to a different lifestyle. Victoriano proudly told us they knew how to live sustainably, respecting the forest with its plants and animals, consuming only what was required. They had no waste. These exact values were shared by Australian (chapter 22) and North American Indigenous peoples. Nutritional research on tribes in this area found that hunter-gatherers eat 130 different species of foods compared with our 63 species. Of these 130 species, 76 were wild fruits, very few were vegetables, and the rest were fish and some meat. It has been suggested that this diet, rich in phytochemicals, may be responsible in part for Amazonian Indigenous peoples' ability to maintain an exceptionally high visual acuity throughout life.[1]

On our second day in Cuyabeno, disaster fell. An oil spill from a nearby well had covered the lakes with a thick, black, stinking film. The Siona, like other groups of Indigenous peoples in this area, depend on the river and lake water for drinking, fishing, and washing. There was a scramble to find a container to collect rainwater and water from less polluted areas up the river, but all they could find was an empty rusty oil drum. As soon as we got back to Quito, Tjitte and Cecilia sent them several water containers.

The oil company, it turned out, had taken a shortcut and not used appropriate settling beds. Oil exploitation had been going on in the area since the 1970s, and this was not the first time there had been an oil spill. The Siona and several other Indigenous groups in the area were suffering from the combined effects of oil spills and illnesses associated with polluted water and unfamiliar viruses, compounded by logging, the building of roads, the drug trade from nearby Colombia, and encroaching agricultural land. Lawsuits against the oil companies were organized but repeatedly failed. Victoriano discussed with Tjitte and Cecilia the possibility of opening the area for ecotourism, in the hope of encouraging people to appreciate the extraordinary richness in biodiversity and the need to conserve this unique area. The effort succeeded, and today there is flourishing ecotourism centered on Siona Lodge, which receives rave reviews. However, episodic oil spills and gas flaring still go on, as does the destruction of the forest.

The Cuyabeno reserve is close to the Yasuní Biosphere Reserve, where the Universidad San Francisco de Quito has a research station on the Tiputini River. We visited this station in 2012. Together, Cuyabeno and Tiputini have the highest diversity of endemic reptiles, amphibians, insects, birds, and bats in South America and possibly the world. In 2007, Ecuador's then president, Rafael Correa, proposed that if the world came up with $3.6 billion, he would leave the oil in the ground. The plan lacked clarity—for example, it was unclear who would receive and control the money. The United Nations and private donors raised $13 million, but having failed to get the full amount, Correa terminated the plan in 2013, and the drilling for oil has continued, as have the oil spills.[2] This disastrous situation lasted until April 26, 2019, when the Waorani people achieved a small success. They won a lawsuit that saved them and nearby peoples from any further oil exploitation on their land. The law says that communities must be consulted before any extraction process is planned on or near their territory. Spurred on by this success, in 2020 they filed another lawsuit, against PetroOriental, claiming that the smoke from gas flaring contaminates their land and threatens their health and survival. We shall see if this too is successful.

Meanwhile, in Tiputini, John Blake and Bette Loiselle, conducting bird surveys from 2001 to 2014, were finding a decline in insectivorous birds.[3] Insects have been declining globally. The main reasons proposed are insecticides, herbicides, urbanization, light pollution, climate change, and habitat degradation; but it was initially surprising that such a strong and significant effect would be seen so deep in the Amazon forest.[4] People who had visited the area told us that the nightly gas flares from the oil companies were contributing to the decline. Flying insects are attracted to the flames, burn, and fall to the ground in such numbers that the oil companies truck piles of them out of the area because of the unpleasant smell.

Unfortunately, the Tiputini situation is not exceptional; the problems of decreasing biodiversity and the impact this has on the lives of Indigenous inhabitants, from whom we have so much to learn, occurs in hot spots of biodiversity throughout the world. A 2022 article in *Science* about similar threats in Madagascar ends: "In the eyes of much of the world, Madagascar's biodiversity is a unique global asset that needs saving; in the daily lives of many of the Malagasy people, it is a rapidly diminishing source of the most basic needs for subsistence. Protecting Madagascar's biodiversity while promoting social development for its people is a matter of utmost urgency."[5] Another, by Jonathan Watts, in the *Guardian*, on the rapid destruction of the Brazilian Amazon forest, states: "To be in the Amazon in 2022 is to live between a tipping point that humanity must avoid and a turning point that we must invent."[6]

22

Australia

> Our spirituality is oneness and an interconnectedness with all that lives and breathes, even with all that does not live or breathe.
>
> —MUDROOROO

Our next opportunity to visit new territory originally occupied by Indigenous peoples came in 2016, when we were invited to be Charles Darwin Scholars at Charles Darwin University in the city of Darwin in the Northern Territory. This was my first and only visit to faraway Australia, and I was delighted to be in the remote north, in an area where there was still easy access to endemic flora and fauna and groups of Indigenous peoples. We stayed in an apartment on campus surrounded by wildlife and with plenty of birds to look at from the balcony: stone-curlews, mound-building megapodes, ibises, Masked Lapwings, honey-creepers, Magpie-larks, Double-barred finches, and Rainbow Bee-eaters. Each evening flying foxes (fruit bats) streamed out, passing us on their way to forage on nectar, pollen, and fruit.

As visiting scholars, we gave several lectures on our research, radio interviews, and contributions to a massive open online course, or MOOC, run by Keith Christian. Keith is a biologist with an interest in the physiological ecology of a variety of organisms, from weaver ants to toads to crocodiles, and a second, very different interest in communities of cyanobacteria that live under transparent quartz crystals in rocks. Keith was a superb companion and excellent teacher, guiding us through the production of the MOOC and introducing us to local wildlife. We had to be satisfied with a bower without the bowerbird that had made it, but we had better success with two other birds, the shy Rainbow Pittas on campus and a remarkably cryptic Tawny Frogmouth (*Podargus strigoides*), almost impossible to find as it looks extraordinarily like the broken branch of a tree.

There were fewer students on campus than I had expected. A large number take their courses online, hence the MOOC, giving them the freedom to work as well as study. Many of the undergraduate students on campus were Indigenous peoples from Australia and New Guinea. The university vice-chancellor, Simon Maddocks, gave us some insights into their lives. He had been brought up in Port Moresby in Papua New Guinea and actively encouraged the interchange between cultures on campus. I could see this in the wonderful displays of Indigenous paintings and information on the peoples' history, art, and music, in a building devoted to this message of tolerant fusion of ideas. Simon told us stories about some of the difficulties experienced by the students from New Guinea when they come to the campus. They had been brought up in communities where people do not own objects but share possessions and, as a result, were sometimes bewildered when their normal behavior was considered stealing in Australia. You could not meet a more empathetic and compassionate person to smooth over the awkward interactions of two very different cultures. We became good friends with Simon Maddocks's assistant, Maryanne McKaige and her husband, Alan Anderson, an expert on ants, and they lent us their tent so we could camp in Kakadu and Litchfield National Parks.

The Kakadu National Park, with its rich biodiversity of indigenous plants and animals, is probably best known for its rock art, the oldest examples dating back twelve thousand years, and some possibly as old as thirty thousand years. The petroglyphs depict animals, such as kangaroos, crocodiles, and fish; and people with digging sticks, dressed in elaborate girdles, armbands, and headdresses, some holding bags and spears. We learned that the tradition is to frequently touch up the original paintings, even to the extent of adding new techniques and new messages, so we should think of them as complex multilayered histories left across the generations to inspire future generations to reflect on life and culture in the past and project into the future.[1]

While at Litchfield, we woke up one night to find an army of Cane Toads marching toward our tent. This was the result of a biocontrol project gone wrong. In 1935, Cane Toads were introduced from Hawaii to Queensland, Australia, to control beetles that were consuming sugarcane. By 1959 the amphibians had spread throughout Queensland, and they reached Kakadu in 2009. The toads are extremely poisonous and kill many animals that attempt to prey on them, including freshwater turtles and crocodiles. Some animals have learned to avoid the toad's toxin—for example, Black Kites will attack the belly, avoiding the poisonous glands situated on the back of the toad's head. Although all life-history stages of the Cane Toad are toxic, a few snakes and the aquatic frog *Litoria dahlii* are resistant to the poison and capable of feeding on the tadpoles. The toads have even influenced natural selection in several species of snakes, resulting in a shift toward smaller mouths, as a result of the larger-jawed individuals dying from eating the large poisonous toads.[2] The army of toads we saw that night made us realize what a gigantic problem eliminating them would be.

Just outside Litchfield, in a swampy area, we found the magnetic mounds built by *Nasutitermes triodiae,* a grass-eating termite species. The mounds are termed "magnetic" because they tend to be aligned on a north–south axis. Some were as high as twenty-five feet, built and configured to ensure that there is a maximum amount of air circulation and cooling for the colony. The mounds are thought to be extremely old, because as the land gets flooded, they are built up and modified to avoid water. In Brazil, the mounds of a similar species have been dated and found to exceed four thousand years in age. It would be interesting to know just how old these ones were.

A brief visit to Alice Springs to give a seminar gave us a chance to visit Uluru (Ayers Rock), the giant dome over a thousand feet high made of a conglomerate of sandstone and feldspar that glows red in the rising and setting sun. When we arrived, rain was pouring down, bouncing off ledges in fast-flowing waterfalls. Michael Misso, director of the Uluru-Kata Tjuta National Park, and his wife kindly drove us on the visitors' road round the rock and explained how exceptionally lucky we were to experience rain on our visit. Rain here is erratic and episodic, arriving only once every few years. Local animals and plants have adapted to the extreme environment. We witnessed frogs that had suddenly emerged in great numbers, having spent the previous four dry years buried deep in the sand. When the percolating water reaches them, they emerge, eat massively, and breed. When the land dries, they fill their bodies with water before digging back down to their subterranean home to wait for the next inundation. Michael explained that before Europeans arrived in the 1870s, there had been forty-six species of native mammals in this area of the park; now there are fewer than half that number, the last survey recording only twenty-one species. Great efforts are being made by him and others to save the survivors, mostly in captivity, along with rare and endemic plants.

The gigantic rock, standing alone on a flat landscape, has long been a sacred site for Indigenous peoples. The local Anangu people have requested that it not be climbed because of its spiritual significance. We were saddened to see that wish disregarded. Iron pegs with ropes had been attached to the lower steep surface, and tourists were being escorted up the rock. Imagine a hoard of tourists descending on Notre-Dame, Saint Paul's Cathedral, or the Vatican with ropes and crampons.

In the following year, 2017, a group of Aboriginal peoples issued the "Uluru Statement from the Heart," a petition asking for greater representation in the Australian government with an advisory body called the First Nations Voice and a "Makarrata Commission." *Makarrata* is a Yolngu word meaning "a process of conflict resolution, peacemaking and justice." Part of the petition reads: "Our Aboriginal and Torres Strait Islander tribes were the first sovereign Nations of the Australian continent and its adjacent islands, and possessed it under our own laws and customs. This our ancestors did, according to the reckoning of our culture, from the Creation, according to the common law from 'time immemorial,' and according to science more than 60,000 years ago. . . . How could it be . . . that peoples possessed a land for sixty millennia and this sacred link disappears from world history in merely the last two hundred years?" One could reflect here on the rate of change to the biodiversity that occurred in sixty thousand years, and the acceleration in change in the past two hundred years as the sacred link disappeared.

This petition was initially rejected by Prime Minister Malcolm Turnbull in 2017. However, it elicited such a groundswell of support that by 2022, Dale Agius, a First Nations man, was appointed commissioner of the First Nations Voice advisory body. His role was to consult with Aborginal peoples throughout Australia to lay the foundations for the implementation of the "Uluru Statement from the Heart." Unfortunately, there was another setback for Indigenous peoples on October 14, 2023, when 65 percent of Australians voted in a referendum to reject a proposal to recognize Indigenous peoples in the constitution. A strong counter-attack raises hope that this decision will be reversed in an election to take place in March 2024. A long uphill battle!

From Uluru we flew to Brisbane and from there drove to O'Reilly's Rainforest Retreat on the Lamington Plateau. A second visit for Peter, it was the first for me. Here was a very different landscape, a dense, lush subtropical rain forest with numerous trails and an extraordinary richness of endemic plants and animals. Many of the endemic trees in this area are the same species that existed in Gondwanan rain forests in the late Cretaceous, as evidenced by fossil trees and pollen cores. Some Antarctic Beeches (*Nothofagus moorei*) were estimated to be well over a thousand years old; they reproduce clonally by suckering, making it difficult to date the original tree. Hoop Pine (*Araucaria cunninghamii*) and Brush Box (*Lophostemon confertus*) were interspersed with Chilean Pine (*Araucaria araucana*), and all were covered with strangler figs and other vines, which made it challenging to spot birds flitting through low foliage or scuttling along the ground. We had quick glimpses of Paradise Riflebirds, whipbirds, and logrunners. The calls of lyrebirds were tantalizingly close, but despite trailing them we never actually saw one.

23

The Diversity of Asia

> Our ability to reach unity in diversity will be the beauty and the test of our civilization.
>
> —MAHATMA GANDHI

International conferences are stimulating, as they allow people to meet, exchange ideas, debate, collaborate, and innovate. In the past, trade brought peoples of different cultures together, and while goods were exchanged, so were new ideas and philosophies. I was not disappointed; our visits to China, Korea, and Japan gave me fascinating insights into the meeting of cultures, the power of a single leader to influence thousands, and the importance of education to awaken an interest in and tolerance of different cultures.

The opportunity to visit China came in the form of an invitation to give a plenary address at the International Ornithological Congress in Beijing in 2002. As is common with many international conferences, our Chinese hosts had arranged several trips for us to visit the highlights and marvels of their country. Peter and I chose to see the Great Wall and the tomb of the terra-cotta warriors and, to satisfy our lifelong wish, to visit Tibet.

The Great Wall, part of the Silk Road, was impressively higher and wider than I had expected from the many photographs I had seen. It was built twenty-six hundred years ago as a fortification to protect traders from warring nomadic tribes from the north. The Silk Road, a conduit between East and West, can best be thought of as a network of connecting trails that extends over thirteen thousand miles. Along these routes, Chinese silk, porcelain, gold, and other articles were transported and traded in exchange for goods brought from as far away as the Baltic states, Italy, Egypt, Greece, Manchuria, Korea, and Japan. Not only were goods traded but music, philosophies, and religions were transmitted, as were diseases, including plague.

The mausoleum of the terra-cotta warriors, built by Emperor Qin, who died in 210 BCE, was a gruesome reminder of the power of a single man. It contains eight thousand life-size terra-cotta figures of warriors, complete with weapons, and five hundred figures of horses and chariots, all arranged in military formation to accompany the emperor into the afterlife. According to Sima Qian, writing a hundred years after Qin's death, all laborers and craftspeople and many of Qin's concubines were buried along with the army. The tomb was sealed, and trees were planted above it to dissuade plunderers.[1] Qin ruled for only eleven years. In that short time, he oversaw the standardization of writing, money, and weights and measures, formed a formidable army, and extended and fortified the Great Wall. He is known for his ruthless oppression of Confucius scholars by burying them alive and enforcing his laws with strict punishments.

What we were seeing in the mausoleum, though unique, had an echo of Western influence. An interesting article by Duan Qingbo points out that everything in the tomb, from craftsmanship to technology, including gold- and ironwork, pan-tiled roofs, and stone sculptures, is distinct from Chinese culture of the time, instead showing similarities with techniques from as far away as Persia and Greece.[2] Qingbo suggests that Qin's political system was strongly influenced by the Persian Empire and Hellenistic kingdoms and that these acted as models for Qin's political revolution. Qingbo raises the question, "What circumstances will actually enable elements to be accepted by a culture, allowing for successful interchange and diffusion?" It is a question I come up against on every continent I have visited. Here we saw evidence of rigid control of people by a single powerful ruler and suppression of opposition.

As if to counteract this display of power and wealth by one man, an exhibit in a building close by depicted life during the early to mid-Tang dynasty, from 618 to 705 CE. This period is considered by many to be the golden age of Chinese cosmopolitan culture. Trade along the Silk Road and by sea was at its height at this time. People from Turkey, Persia, Constantinople, India, Korea, and Japan were in contact with each other. The exhibit showed poetry, painting, and music, with imported instruments resembling oboes, flutes, drums, and cymbals. Different religions were tolerated; the role of women in society changed, and Wu Zetian became the first and only empress in Chinese history. Taoism, the official religion, emphasized living in harmony with nature. Eventually the dynasty ended, in 907 CE, when a massive flood followed by drought caused crop failure, widespread starvation, and death. People rebelled against their leaders and overthrew them. The rise and fall of a powerful empress and emperor in connection with cultural exchange made me reflect on Luigi Luca Cavalli-Sforza's research and writings discussing how rapid changes in society come about when one dictator has the power to transmit his or her ideas to thousands.[3] The question then becomes, what elements of societies permit a leader to be benevolent or malevolent? Would we gain insights into this question in Tibet, where the Buddhist religion is considered to epitomize tolerance?

Peter and I had always wanted to visit Tibet, but it had seemed out of reach until now. It was a dream since school days, when my geography teacher lent me her copy of *Seven Years in Tibet* by Heinrich Harrer.[4] Peter had also read this book as a boy. Harrer was an Austrian mountaineer who had escaped from a prisoner-of-war camp in India and hiked over the Himalayas to Tibet. There he met the eleven-year-old Dalai Lama and taught him English, history, and mathematics, while relaying the BBC news to the Tibetan government. The book is an outsider's description of life in Tibet from 1944 to 1951.

Our Chinese guide came down with altitude sickness, and a Tibetan guide substituted for him. He made sure we experienced Tibetan food and music, and even invited us to take part in a lively traditional country dance one evening. In Lhasa's central market, human skulls were sold at one stall, elongated wooden butter churns at the next, and Tibetan clothes at a third. Little children with chubby red cheeks rushed about, crawling underneath tables and playing tag. They were fascinated by my white hair and took it in turns to touch it and look through my binoculars. Looming over Lhasa and all this activity is the extraordinary Potala Palace, a huge fortress built into the rock and home to a succession of Dalai Lamas from 1649 to 1959. We visited it one morning with our guide, first the library, holding numerous scrolls, then through halls and shrines of past Dalai Lamas, all decorated with paintings and elaborate carvings.

To see a small part of the local countryside, we joined a bus tour to a woman's monastery (or nunnery) outside Lhasa. The bus took us along a tributary of the Brahmaputra River, which eventually joins the Ganges River delta. People were fishing from circular boats, or coracles, just like ones used for fishing in Wales, England's Lake District, and Scotland. We passed fields of barley, the staple cereal crop, which is dried and roasted, then ground into *tsampa*, a flour that finds its way into everything, often mixed with yak butter. Lammergeiers, or Bearded Vultures (*Gypaetus barbatus*), circled overhead waiting for a sky burial. After the spirit has left the human body, this now mere vessel for the soul is cut into pieces, sprinkled with *tsampa*, and taken to the top of a mountain to be devoured by Lammergeiers and recycled into life.

After several hours, we reached the foot of a rocky mountain covered in small shrubs. We climbed slowly, feeling the effects of the altitude, up a long, stony path with two pairs of Tibetan Eared Pheasants and their chicks scuttling in front of us. At the monastery, the only sign of human life was a group of nuns incongruously peeling potatoes before dropping them into a huge caldron—potatoes that had originated in the Andes and been brought along the Silk Road in the nineteenth century to Tibet, to thrive in another alpine environment. The wooden table and caldron of potatoes reminded me of an evocative, unusually realistic painting by a young Vincent van Gogh of poor women spearing boiled potatoes from a caldron before eating them.[5]

We were not in Tibet long enough to fully appreciate the interaction of different cultures. Visits to Korea and Japan gave us a deeper understanding of the impact of communication between cultures and the effect that present-day education has in awaking tolerant curiosity to new cultures.

For almost two thousand years, Korea was a conduit for trade and culture between China and Japan. Two visits to Korea enabled us to appreciate how much Korean architecture, art, music, and religion had been influenced by ancient Manchurian and Chinese culture and how it had become creatively modified over the centuries into distinctive customs and ways of life. The influence extends to writing. It was explained to us that the same Chinese characters are used, the difference being that in China these characters denote sounds, whereas in Korea and Japan they represent the alphabet.

Both times we visited South Korea our host was Jae Choe, whom we had met in the United States. He had studied at Harvard before returning to Korea as professor of biological sciences at Seoul National University and later at Ewha Womans University. On our first visit, in 2002, Peter was to give the keynote address at the International Association for Ecology (INTECOL) meeting, and I was asked to give a short, ten-minute predinner speech on what it was like to live and do research on Daphne Island. The dinner was an exalted occasion, with Jae escorting the mayor of Seoul and the governor of the province to the table. Something had gone wrong, and as I was walking to the stage, Jae rushed toward me and whispered, "*Please, please,* can you turn your ten-minute talk into forty minutes?" Somehow, I managed to spin out the slides with stories and finish exactly forty minutes later, on the dot. I still do not know what the delay was about, or how I managed this feat, but I am glad to have been able to repay Jae's generosity in inviting us to South Korea in the first place.

Jae provided us with five lively graduate students and a van. They took us to Changdeokgung Palace, the National Arboretum, and a Korean folk village before treating us to a traditional Korean lunch. This entailed cooking strips of meat on a small barbecue grill placed on the table. We then dipped the sizzling meat into bowls of colorful spices before wrapping it in sesame leaves. The barbecue grill was identical to a two-thousand-year-old grill we saw in a museum in Xi'an, China!

The day ended with a superb musical concert played on traditional instruments at the Seoul Arts Center. Here we experienced the strong influence of creative communication between cultures in both the musical instruments and the music itself. Some instruments resembled zithers, with varying numbers of up to twenty-five silk strings; others were forms of lutes, bamboo flutes, oboes, drums, and gongs, played together or separately. Their counterparts can be found in China and India and, like the alphabet and the music they played, creatively modified.

Our second visit to Korea was in 2015. This was a very special and humbling occasion. We were invited to inaugurate a trail labeled Darwin's Way and a companion one labeled Grant's Way at the new National Institute of Ecology property located near Gunsan. We walked along Darwin's path, a Korean version of the famous sand walk at Darwin's home in England. This one wound uphill through deciduous woodland into a stand of Korean Pines. At intervals, there were slabs of polished black stone with significant events in Darwin's life inscribed in white, in English and Korean. Grant's Way forked off this path. The first marker of our research career was a hut closely resembling our Daphne cave, complete with metal storage boxes and simulated *chimbuzos* (water containers). This was followed by other markers referring to our research. The trail eventually descended to rejoin Darwin's Way, leading to the final Darwin marker, bearing a nice tribute composed by Jae Choe: *Darwin's greatest contribution to humanity was to humble us, as a species, once and for all.* The unveiling of a plaque took place in front of a circular pool with islands representing the Galápagos archipelago, again creatively designed by Jae. Both Peter and I felt overwhelmed by this honor and attention.

The National Institute of Ecology, another brainchild of Jae Choe, illustrates the power of hands-on education in awakening curiosity. The institute's mission is: *Contribution to the realization of the sustainable future by promoting research and conservation of the natural environment and spreading eco-culture.* We were curious to see how this was achieved.

Five large buildings are dedicated to five major biomes: tropical, desert, Mediterranean, temperate, and polar. In each, visitors can follow various paths meandering between plants and animals typical of the region and experience the differences in temperature, humidity, and smells among them. Care had been taken to subdivide each biome into distinct areas, with explanations given for their similarities and uniqueness. For example, the tropical biome was divided into African, Asian, and South American areas. The education section in another building had classes for elementary to high school students, lectures covering career paths and jobs, and advertisements for outdoor camps. While we were there, some parents with excited young children brought in ants they had found. The children were shown how to identify them and then set up experiments to test their own questions. The experiments could be left running in the laboratory until the children returned some days later. Jae had found that such activities excited not only children but also their parents, many of whom were interested in and asked questions about nature for the first time in their lives. Both parents and children were becoming passionate about the need to conserve the global environment. The program has deservedly received international recognition.

On our way back to Seoul with our guides, Gilsang Jeong and Jongmin Kim, we stopped at two temples. Notable at a cluster of temples at Naesosa was a venerable thousand-year-old elm-like tree (*Zelkova serrata*), looking huge, ancient, bent, and crippled. A fifteen-hundred-year-old temple at Seonunsa that was surrounded by camelias epitomized tranquility and invited reflection on our recent experiences.

We gave lectures at two private universities, Konkuk and Ewha, the latter a university for women. Here we had long discussions with students about their own research aspirations. They were just as keen to discuss how to balance family life with a career, a topic that clearly concerned them as they moved from a structured to a more open, flexible, and egalitarian society. Change was in the air.

Before leaving Korea, we spent a couple of interesting hours at a comprehensive cultural museum. On the grounds were small wooden totem poles, an unsuspected link to Canada and the movement of people across the Bering Strait. Korea appeared to have acted as a hub, with a spoke to China, another to India, a third to Japan, and another to Canada.

We visited Japan on three wonderful occasions, in 2003, 2009, and 2018. In 2003, we participated in scientific meetings in Tsukuba and Kyoto. Kyoto is like nowhere else, located in a valley surrounded by mountains with many ancient temples and shrines. The Japanese gardens that emphasize and enhance the natural landscape were ubiquitous, we found, from those surrounding temples to tiny gardens of natural moss, stones, and an azalea or camelia tastefully arranged in the smallest courtyard. An atmosphere of permanent, undisturbed peacefulness accompanied us as we walked up through a sunlit forest with a light dusting of snow toward Enryaku-ji, a temple at the summit of Mount Hiei. We were jolted out of this tranquility on reaching the first sign, which told us of a turbulent, warring history and how the temple had housed a huge army of warrior monks that aggressively defended the area until they were finally defeated in 1571.

Our visit in 2009 was a once-in-a-lifetime experience. Peter and I had the very special and moving honor of being jointly awarded the Kyoto Prize in Basic Sciences from the Inamori Foundation. It is considered Japan's most prestigious award. In the words of Kazuo Inamori, it is an international award "to honor those who have contributed significantly to the scientific, cultural, and spiritual betterment of mankind." Our fellow prizewinners that year were music composer Pierre Boulez in the Arts and Philosophy category and Isamu Akasaki for his invention of the blue LED light diodes in the Advanced Technology category. We were the first joint husband-and-wife recipients of the prize; I hope there will be more in the future. Nicola and Thalia and their children were to have joined us, but sadly, for reasons of other commitments and illness, they had to cancel. However, we did have some family members: my brother John and his wife, Moira, and Ravi, Nicola's husband. Powered by adrenalin, we were kept busy for eight days that went by in a whirl under the watchful eye of Tohru Suzuki, our guide throughout the week. Like an attentive sheepdog, he guided us through every minute step of the way, but more than that he was a knowledgeable companion, being an agricultural scientist, and

FIGURE 17. Upper: With Peter receiving the Kyoto Prize, 2009. Middle: Peter and Dr. Tohru Suzuki at Koke-dera (Moss Temple), Japan. Lower: Golden Pavilion at Kinkaku-ji, Kyoto.

fully informed about not only science but Japanese culture and its historical significance.

The awards ceremony took place on the day after our arrival in the Kyoto International Conference Center. A full orchestra occupied a balcony above the central stage, which was draped with a blue-silk curtain. There were strict instructions for me to wear a long dress that must not be black, and to the delight of the organizers, I had serendipitously chosen a silk dress of exactly the same color as the curtain.

Interspersed between intervals of orchestral music were speeches and a Noh play. The British ambassador read a message of congratulations from UK prime minister Gordon Brown. Then, after Dr. Kazuo Inamori and Princess Takamado's speeches, we each gave our short acceptance speeches. The ceremony ended with a chorus of young children who presented us with a traditional Japanese embroidered ball.

The banquet that followed was magnificent. About one thousand guests were seated at parallel tables arranged perpendicular to our table, which was on a slightly elevated platform. We laureates sat in the middle between Dr. Inamori and Princess Takamado, flanked by dignitaries from the city. Communication took place through impressive interpreters who sat behind our shoulders. After some short speeches, the banquet ended, and we followed the princess to the ballroom. As the younger people began to dance, we sat down with the princess at a table. She told us she wrote children's stories, and we found that she was a knowledgeable and keen bird-watcher. Later, I bought several of her delightful and beautifully illustrated children's books to give as presents.

Dr. Inamori valued the importance of education at all levels. During the following days, we not only participated in symposia with other professors but interacted with schoolchildren both at high school and elementary school levels. The first symposium was devoted to evolution. Our lecture on Darwin's finches was followed by one given by Yoh Iwasa, a brilliant mathematician interested in many biological problems, from mate preferences to cancer; a talk on nest-parasitic cuckoos by Hiroshi Nakamura; one on cichlid fish in Lake Victoria by Norihiro Okada; and two lectures on plant-animal interactions by Tetsukazu Yahara and Makoto Kato. We paid two visits to schools, the first to a group of high school students in a large hall with girls on one side and boys on the other. It began with a tea ceremony performed by two boys. We gave a short talk and then took questions. Peter had a hard time hearing the questions, and when a young boy apologized for his poor English, the translator explained, "No, don't apologize. Your English is excellent; he is just an old man with poor hearing!" That broke the ice, everyone laughed, and the questions came loud and fast.

The second visit took us to an elementary school for a "kids' science" session. We had taken some tweezers and pliers of different sizes and shapes, along with seeds and nuts, representing beaks and food of finches. Everyone in the room wanted to try cracking the seeds, and the collective enthusiasm overwhelmed any attempt we made at controlling the activities. A young boy took the smallest sharp-nosed tweezers and successfully cracked the largest nut. We quickly turned a negative into a positive amid much laughter by declaring how this shows that beak shape is not everything—jaw muscles also matter! A small boy asked, "How old were you when you got married?" "Twenty-five." "Ooh, that's young!" came a chorus from the room. Then another: "What do you like about each other?"

Another entire day was devoted to press interviews. Clearly, it was unusual for us, as a husband-and-wife team, to be conducting joint research, and there were still relatively few women scientists. Many of the questions were related to how we worked together and at the same time raised two children. This was a more sophisticated version of the same question we got from the children. At the end of the day, one of our interviewers bowed, then warmly shook my hand and said, "I am going home to tell my wife that from now on I will be very considerate to her!" He was such a gentle, pleasant person that I am sure he was considerate to her before meeting us, but the message of the joy of cooperation went home.

Tohru escorted us to the surrounding temples on our last day. We first visited the famous Moss Temple, Koke-dera, and listened to monks chanting, then walked along small winding trails past miniature waterfalls and ponds through the gardens. Reputedly, 120 different species of moss form the carpet under the trees, now in glorious autumn colors. Several monks in soft boots were sweeping the fallen leaves from the moss with soft brooms. The Ryoan-ji Temple, with a Zen stone garden of seven rocks surrounded by a sea of gravel raked into a different pattern each day, reminded us of the tranquility that can be experienced by simply staring at irregular patterns of stones. We hurried to see the temple of Kinkaku-ji at the perfect moment as the last rays of the sun fell on the Golden Pavilion; and then Nijo-jo castle, where the shoguns resided during the Edo period. The castle's floors made a chirping sound when we walked, designed to alert the guards that someone was about. Most spectacular were the paintings on the doors and walls of pine trees and birds. I particularly remembered an eagle and goshawk. Unlike Western paintings, they had no perspective of depth or distance. Nevertheless, the similarity to the arrangements of pine trees in the garden outside was striking, and without perspective, it gave me the uncanny impression of continuity of inside and outside, as though there were no walls in between.

In 2018 we returned to Japan, this time to Sapporo for a conference on the mechanisms of evolution and biodiversity. We arrived at night and gave our joint plenary address the next day, having not seen any natural daylight for two days. The effect was, as Peter described, like giving a well-rehearsed lecture while sleepwalking. I realized that part of our problem was lack of light. From then on, at lunchtime we went into the outside world for a walk. There had been a very large earthquake the week before, which hit the headlines of all the international papers, and we had to climb over fallen trees and jump over cracks in the pavement on the trail around the hotel. One day, as we sat listening to a lecture, alarms went off from every computer in the room, blaring "earthquake alert" in Japanese and English. It was a 5.3 magnitude earthquake—only a small aftershock, said our host. The luxurious Châteraisé Gateaux Kingdom hotel, where we were, swayed but stayed grounded. The stable mountains of Hokkaido, seen in the distance through the window, tantalizingly beckoned us, and we would have loved to have extended our stay to hike and explore them, in the light and on stable ground.

24

Return to Europe

We shall not cease from exploration,
and the end of all our exploring
will be to arrive where we started
and know the place for the first time.

—T. S. ELIOT, *FOUR QUARTETS*

We took every opportunity we could to return to Europe, and I had the chance to do so almost every year through invitations to seminars, teaching, and being on a committee of the Royal Society. I felt always as though I was coming home and seeing the place for the first time. On our visits to different countries for university seminars and other collaborations, we made a practice of delving into the history of the country, visiting art galleries and museums, taking long walks, and observing the local natural history. Were there parallels in European and Asian history and commonalities in the causes for periods of oppression versus times of tolerance and great creativity? Asia had given us fascinating insights into the meetings of cultures, the power of a single leader to influence thousands, and the importance of education to awaken an interest in and tolerance of different cultures.

Apart from Britain, the regions that were most meaningful to me were Scandinavia, Switzerland, and the Mediterranean countries. Scandinavia is special. My first trip out of Britain was to Norway when I was sixteen. Sweden was where I received my PhD (chapter 15), and over the past ten years, Peter and I have been collaborating intensely with Leif Andersson and his students at Uppsala University on our research (chapter 16). Finland inspired me with its education system (chapter 14). Iceland was where the charr lived that I had hoped to study (chapter 8) and homeland of the relatives of the Herdwick sheep and the people who had visited Newfoundland long before Columbus's day (chapter 6).

In 2009, we returned to Norway, where I gave the Kristine Bonnevie Distinguished Lecture at the University of Oslo. I felt a kinship with Bonnevie, who was sixty-four years older than me to the day. Not only did we share the same birthday, but her father, like mine, was opposed to women in academia. She eventually overcame that hurdle to study the same subjects as I did: embryology, genetics, and cytology. Her research was on the subject of heritable abnormalities such as polydactyly. I was interested to learn that she was supported in her role as a scientist by men more than other women. Two biologists, Georg Sars and Robert Collett, persuaded the Norwegian parliament to pass an act in 1912 to grant women the same right as men to hold the position of professor at a university. Another outstanding offer of aid by men in power was given to Caroline Herschel, who lived from 1750 to 1848. She was William Herschel's sister and an outstanding astronomer in her own right, having discovered several comets. She received continuous encouragement from Jérôme Lalande and Nevil Maskelyne, the British astronomer royal, for her scientific research. She was the first woman to publish in the *Philosophical Transactions of the Royal Society*, and toward the end of her life, at age ninety-six, she received the Gold Medal for Science from the king of Prussia, conveyed to her by Alexander von Humboldt.[1]

These two examples resonated with me. In my own career, I had been helped at turning points by sympathetic men, usually men with daughters of their own. I have also been helped by women; Charlotte Auerbach was an outstanding example when I was an undergraduate. But there have also been occasions when I have been treated with disdain by women, which I interpreted, given the context, as a kind of social competition. In the 1950s, this was well known among us female students as the "Queen Bee effect." Fortunately, with the increase in women in professional situations, I believe this is no longer the case, and the Queen Bee has outlived her life. Frequently, I am asked to join or hold women's meetings to discuss their difficulties in pursuing a career in a man's world. Too often these turn into sessions of (justifiable) complaints. It is often considered the role of the oppressed to assert their rights. I try to persuade such groups to include people of all genders in the meeting and to particularly include those who have the power to overturn entrenched institutionalized inequalities. Very often, those in power have been ignorant or blind to such inequalities, and they often start by defending the current situation but quickly reverse their original opinion in such meetings of minds. Communication is the key to resolving misunderstandings, and when successful, it will rapidly provide much-needed support for all who are disadvantaged, be it through gender, race, poverty, or other circumstances.

Ten years later, two seminar invitations took us to Finland. In Helsinki, along with Jared Diamond and Dieter Ebert, I had been asked to give a talk in honor of Ilkka Hanski, who had recently died. Peter and I admired Ilkka and his research. His research was similar to ours but with interesting differences that made discussing the two together enjoyable and compelling. Ilkka had studied the Glanville Fritillary (*Melitaea cinxia*) in a network of four thousand small meadows in the Åland islands. Each year about one hundred of the four hundred to eight hundred local populations of these butterflies go extinct through the drying out of vegetation, parasitism by wasps, and inbreeding depression. Yet, this network persists because of immigration of colonists from other meadows that differ genetically and physiologically from the extinct populations. The importance of spatial heterogeneity in the butterfly populations contrasts interestingly with the temporal changes combined with episodic introgressive hybridization in the finches. In both cases, variation, in phenotype and genetics on which selection acts, is produced and maintained.[2]

From Helsinki we took a train north to Jyväskylä, where Johanna Mappes had invited both Peter and me to give seminars at Jyväskylä University. Johanna is an evolutionary ecologist specializing in interspecific interaction, in the field and with experiments in the laboratory. The day after our seminar, she took us out into the country with a few keen bird-watchers who knew where to find two enormous owls, a Ural Owl and a Great Gray Owl. We ended the excursion with a scramble down a cliff to see the Saraakallio rock painting. What a wonderful day it was, though cold, and as I looked up at the paintings, done sixty-six hundred years ago, I felt that the last ice age in Finland was not very long ago. There, many feet above us on a cliff overhang, were etches of a realistic Moose, deer, and a Viking-style boat with several humans in it, all painted with a mixture of hematite soil, blood, urea, and eggs. As they were so high up, we wondered at first how they could have been painted, until we realized that all those years ago, the lake would have been higher, and they could easily have been done from a boat.

Peter and I went to Finland again in 2022, and on arrival were whisked off to the European Meeting for PhD Students in Evolutionary Biology (EMPSEB), an annual conference that was being hosted by Finland. I enjoy these conferences enormously; we had been to one in Spain a few years before. About fifty young PhD students from all over the world come together for a week to give talks and discuss their interests with peers and professors. Ideas are exchanged, sometimes leading to collaborations and lasting friendships. Interacting with young enthusiastic scientists and seeing the joy they have in exchanging ideas with each other at an early stage in their career is valuable compensation for relinquishing teaching on retirement.

Switzerland, like Sweden, has always had a special place in our hearts. We take every opportunity we can to visit, and there have been many. I was a visiting professor for two successive years of three months each at the University of Zürich, and on many occasions Peter and I contributed to the internationally popular evolutionary biology course in Guarda organized and designed by Sebastian Bonhoeffer (ETH Zürich) and Dieter Ebert (University of Basel). International students in the early stages of PhD studies come together for a week in the Alpine village of Guarda. In groups of four, the students have the task of developing an original biological question and a way to test it. They then write a research proposal, unlimited by funds. After daily discussions with us four or five professors, they give presentations to the class on their research proposals. Some of these proposals have the germs of outstanding projects and have been followed through later with genuine research.

Our trips to Zürich often combine stimulating interactions with colleagues and visits to opera and concerts with time spent hiking in the mountains. One outstanding musical occasion at a moving honorary degree ceremony was a student playing a beautiful interpretation of a piano piece by Mozart. He had been trained through El Sistema, a program founded by José Antonio Abreu in Venezuela, which gives the poorest children free classical music education and access to instruments. I had been fascinated for a long time by this program because its main goal, as described by Abreu, is to provide "music for social change." It is driven by Abreu's philosophy that playing music in an orchestra requires discipline, respect for others, and a spirit of cooperation that is emblematic of what makes us all better citizens of the world. The program became controversial because of its close association with the Venezuelan government. In spite of this, countries around the world, particularly European nations as well as the United States and Canada, have adopted similar programs to introduce young children to classical music, inspired by Abreu's philosophy. Outstanding musicians and conductors of major orchestras, such as Diego Matheuz of Teatro La Fenice in Venice and Gustavo Dudamel of the Los Angeles Philharmonic Orchestra, had their early training in the El Sistema method.

For several summers we spent three weeks in a small chalet that we rented in the tiny village of Jeizinen in the Valais canton. It is reached by either a long trek from Gampel or by Seilbahn, a mountain cable car. Jeizinen lies on the south-facing slope of the Rhône valley with magnificent views of the nearly fifteen-thousand-foot-high Weisshorn. Often we wrote in the morning, then took long, five- or six-hour hikes up into Alpine meadows filled with flowers and butterflies. A glass or two of local wine with dinner ended the day as the sun was setting. The simple things in life are the best!

While hunter-gatherers were painting Moose on rocks in Finland sixty-six hundred years ago, two thousand miles south, agriculture was flourishing, as revealed by pollen grains of wheat, barley, peas, flax, and poppy appearing in core samples. The difference between the younger, far northern societies and the older southern Mediterranean ones from the same era are substantial. On the second of two rewarding seminar visits to Portugal, our hosts, in Porto, were Paulo Alexandrino, Paulo Alves, and Nuno Ferrand. Like us, they are interested in genetic diversity of natural populations. One discussion touched on the early human inhabitants of Portugal, so on our day off, to be spent exploring the countryside, Paulo Alves drove us to a lookout point for viewing a megalithic site on the banks of the Douro River. On the way he pointed out numerous interesting facts, including how the planting of the invasive eucalyptus trees had displaced the local chestnut, Carob, pines, and the Cork Oak. Unlike the roots of native trees, eucalyptus roots do not burn, and these trees are the first to recolonize an area following a major fire and are extremely difficult to destroy—the result of a long history of selection for fire-resistance in Australia. We passed one of the few remaining Cork Oak forests and learned that the bark of these magnificent ancient oaks is stripped every nine years for the cork industry.

In late afternoon we reached the Côa Museum, standing on a bluff where the Côa River meets the Douro. Unfortunately, we had forgotten it was Monday, and the museum was closed.

Made of dark schist, the building appeared to be yet another giant boulder on the sparse landscape. From here we looked down at the far bank of the Douro River and the megalithic site. Dark thunderclouds were rolling in from the west, with flashes of forked lightening, illuminating the scene in a way that helped to transport us back in time. We imagined the ancient people, now known from their genetics to be dark-skinned with light eyes,[3] rowing up the river to shelter in caves where their etched outlines of deer, horses, and aurochs can still be seen. These were the ancestors of the people who first arrived in western Britain and moved into the Lake District. I wondered, did they count in the Westmorland way?

———

Spain gave us insights into another, more recent period in history and different rewards. I had read about Antoni Gaudí's architecture inspired by nature but was unprepared for the extent it infiltrates and dominates parts of the city of Barcelona. Roofs look like magnolia leaves; windows resemble honeycombs and diatoms; columns are like tree trunks wrapped around by vines or perhaps simulations of stalactites and stalagmites from nearby limestone caverns. Even the sidewalks are covered with Gaudí's hexagonal tiles decorated with exotic sea creatures. Was he inspired by the fossils found in the Jurassic limestone plateau known as Les Causses? Its caves include Aven Armand, a massive cavern measuring 330 by 180 feet, within which small paths wind between immense stalagmites, one as high as a hundred feet. Gaudí's cathedral, the Sagrada Família, would not look out of place inside.

The CosmoCaixa natural history museum, perched high up and overlooking the city, was spectacular. The centerpiece of this hands-on museum for children and adults is a tall Ceiba tree from the Amazon rain forest, surrounded by a spiral staircase from the canopy down to the roots in a flooded forest. The exhibit includes over one hundred living animals, from tropical birds, frogs, and alligators to piranhas. In another hall, a stuffed, lifelike replica of a dwarf goat, *Myotragus*, found as a fossil on Majorca, caught my eye. We had recently taken part in a conference in Minorca, where there had been a lively discussion of gigantism and dwarfism on islands, centered on food availability or lack of it and predators acting as selection agents. *Myotragus*, the specimen in front of us, was particularly interesting because microscopic examination of the bones revealed extremely slow and intermittent growth, more like that of a reptile than a mammal. In contrast to this tiny goat, and dwarf elephants found in Crete and Sardinia, giant fossil rabbits (*Nuralegus rex*) have been unearthed in Minorca, and a giant rat (*Canariomys bravoi*) and a giant lizard (*Gallotia goliath*) have been found on Tenerife. All these animals were present from the Pleistocene up to the arrival of humans, when they rapidly became extinct, probably due to a combination of hunting by humans and vegetation changes as global temperatures rose,[4] another indication of how humans have impacted nature.

A seminar took us to Granada. We stayed at Carmen de la Victoria, a hotel owned by the University of Granada, situated directly across the ravine from the Alhambra. Arriving just before sunset, we watched entranced as this palace-fortress, a monument to Islamic architecture, changed color gradually from a somber gray to a brilliant glowing red. On our previous visit we had joined a small group of mathematicians to visit the Alhambra at night. The palace was almost empty as twenty of us were guided with hushed voices from room to room. The smells of Mediterranean herbs filled the air. The trickling of water from the many fountains was audible in the stillness of the night. The fountains are part of a continuous supply of pure water to the fortress, fed through a network of ancient culverts down a natural gradient from streams in the Sierra Nevada. The walls, columns, and arches in many rooms, which were covered in Nasrid cursive script, appeared gray in the moonlight, so we were astonished to see how colorful the walls really were when we revisited the same rooms in daylight the following day. We were told many of the texts were Quranic poems that had been composed especially for the palace walls. Some referred to the beauty of the surrounding nature, while others were in the first person, as if the walls themselves were speaking. I would have loved to have read a translation of some of those poems.

In Granada, mutual tolerance prevailed for centuries among Jews, Muslims, and Christians, we were told. We walked up the slopes of the Albaicín, the old Arab quarter, and through the Realejo, the old Jewish quarter, experiencing how physically close together they were. All broad-mindedness and exchange of ideas ended suddenly in 1478, when Muhammad XII of Granada surrendered Alhambra to the powerful Catholics and fled to North Africa. King Ferdinand II and Queen Isabella I moved into the Alhambra and ordered that Jews and Muslims must convert to Catholicism or leave. The punishing Spanish Inquisition became a dominant force in Europe for more than two hundred years, a dark era, with much literature and art depicting the horrors of those times. The speed of the apparent turnover was astonishing, reinforcing once again, as we had seen in China (chapter 23), how the power of a leader can control many, leading to intellectual enlightenment or darkness.

In the north of Spain we saw signs telling us of an open-minded thirteenth-century king, Alfonso X, who had encouraged communication between members of different religions and arranged

FIGURE 18. Upper: Alhambra, Granada, Spain. Lower: View of houses below the Alhambra.

for Arabic and Latin documents to be translated into Castilian, the vernacular language. This act is considered to have been pivotal in stimulating future Spanish literature, science, and philosophical thought. I thought back to Duan Qingbo's question regarding Emperor Qin's political domination as depicted by the mausoleum of the terra-cotta warriors: "What circumstances will actually enable elements to be accepted by a culture, allowing for successful interchange and diffusion?" (chapter 23).

Tolerance is not the only ingredient of social harmony: communication, cooperation, and access to education that inspires creativity are others. There are interesting parallels in biology. Enrico Coen points out that humans have a fine balance of competitive and cooperative drives and that these are two of the principles applied to the transformation from simple cells to civilizations.[5] As we have seen throughout our travels, this fine balance is fragile and easily perturbed. After our visit to Granada, we enjoyed a few days in Seville at the invitation of Jordi Bascompte, who was applying network theory to ecological communities to examine the role of interactions that yield mutual benefits. Recently, Jordi and his colleague published a paper in which they extend network theory to the importance of medical knowledge held by different groups of Indigenous peoples. Their message is that without communication between people of different languages, valuable information on the medical properties of plants is in danger of being lost.[6] It is, as Abreu indicated in his idea of music for social change, respect and cooperation that will make us all more responsible citizens of this world.

An opportunity to mingle science with art occurred in a visit to the south of France at the Tour du Valat research station in the

Camargue, where we were hosted by Luc Hoffmann, who had also cofounded the World Wildlife Fund. Luc had a unique long-term vision, a commitment to conservation of birds combined with preservation of a rural life. Biologically, the area is interesting because its marshlands contain a gradient from fresh to brackish to salt water. Luc and his son, André, were hosting a concurrent meeting of Van Gogh's relatives, who were in Arles in connection with the Van Gogh foundation. It was our good fortune to be invited to dinner at Luc Hoffmann's daughter's restaurant and to sit next to Van Gogh's brother's grandson. Lively discussions of Van Gogh's life seen through their more knowledgeable eyes sent Peter and me walking round Arles the next day to look at the same scenes that Van Gogh had painted and see a glimpse into this complex character.

Venice, this wonderful city of canals, bridges, art, and music, the only transport by gondolas and water taxis. We were told you must get lost to know Venice. This we did many times in the maze of buildings, bridges, and canals, finding art museums, a museum of old musical instruments, and an exhibit of Leonardo da Vinci's amazing inventions along the way. We gave talks on two occasions in the Istituto Veneto di Scienze. We stayed in one of the institute's buildings next to the church of San Vidal. Each night we listened to concerts of Baroque music, mainly Vivaldi and Handel. Davide Amadio was the leader and cellist. His flamboyant personality set a lively pace to well-known music, for us a new but not unpleasant interpretation.

Giorgio Bernardi, our host in Venice, later invited us to an international genomics and evolution workshop in Naples. We could not resist staying on a few extra days to visit Pompeii,

Herculaneum, and Paestum. Peering into the Vesuvius crater, we could see a few fumaroles actively belching sulfurous steam, reminding us it was only dormant, not dead. It brought the demise of Pompeii and Herculaneum a little closer, a feeling that the destruction had not occurred that long ago. When we walked through the remains of these cities, it was easy to imagine living there, visiting houses, eating at local stalls, jumping over the open drains in the center of the streets, and feel how in a flash all this was instantaneously demolished.

Paestum was very different; it was as though we had suddenly been transported back into Greece. Walking on grass between the ruins of three Greek temples, immersed in the scents of sun-warmed mint, fennel, and thyme, I tried to imagine life in the sixth century BCE. At that time, Paestum had been a Greek colony in Italy, involved in trade with Carthage and other Mediterranean ports. For me, the highlight of this evocative site is the Tomb of the Diver in the museum. It beautifully depicts a diver in midair, about to enter the water, possibly a metaphor for diving into the underworld, the unknown. It has been dated to about 480 BCE, but nothing is known about it. The fresco is unique; no other Greek fresco has such realistic imagery. In a paper describing the tomb, Ross Holloway mentions the similarity in technique with Etruscan fresco painting and also suggests that the fresco could be the work of a local artist inspired by paintings on Attic vases.[7]

The Etruscan link resonated with me. We had visited the National Etruscan Museum in Rome and were impressed by the beautifully proportioned design and frescoes. The latter, often from tombs, depicted people dancing, playing musical instruments, and feasting. The Etruscan civilization had reached its height in the eighth to third centuries BCE and, based on objects found mostly in tombs, was actively trading throughout the Mediterranean, particularly with Egypt, Phoenicia, Greece, and even the Baltic, as inferred from the presence of Baltic amber. The Etruscans were highly skilled in metallurgy, hydraulic engineering, and temple design and had a strong influence on Roman culture. In the first century BCE, Diodorus Siculus, a Greek historian, wrote that literary culture was one of the great achievements of the Etruscans, although today we have no inkling of it. The Etruscan language itself is unique, like Basque, and there are many speculations about its origin. Here is yet another mystery, like Linear A, the Minoan writing that still needs to be deciphered. Clearly, the Etruscans were a sophisticated people open to collaborating with their neighbors. If a time machine could take us back to their time, we might find Etruscan influences at every place we visited to give lectures that are mentioned in this chapter. An interesting thought on the transience of civilizations!

25

Where Do We Go from Here?

> Education is the most powerful weapon which you can use to change the world.
>
> —NELSON MANDELA

Late one evening, Peter and I were sitting on the beach on the uninhabited Galápagos island of Genovesa. The stars, brilliant in a moonless sky, were accentuated by the blackness of the night; waves lapped gently against the shore. Suddenly, a flaming meteorite plunged into the ocean, illuminating for a few brief seconds the abundance of life around us in vivid color. A human life is but a flash in geological time, yet long enough for me to become fascinated by earth's biological and cultural varieties and to explore deeply questions on how and why such a rich diversity of life came to be.

My span of life unfolded at a time of unprecedented change. It overlapped two epochs, the Holocene and the Anthropocene. The boundary between them is situated in the 1950s, when microplastics, heavy metals, carbon particles from fossil-fuel combus-

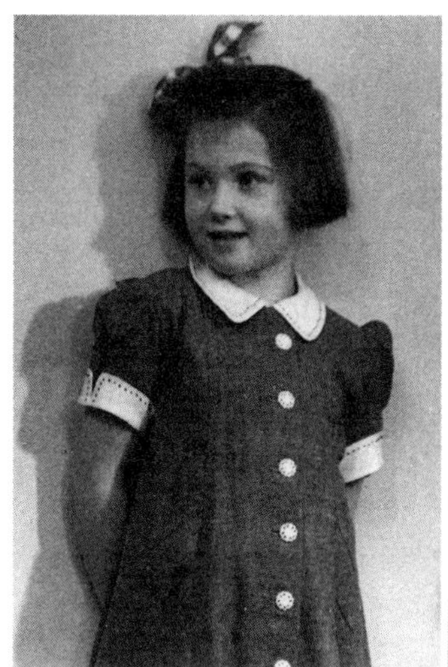

FIGURE 19. Left: Author at about three years old. Right: At eighty-four years old.

tion, and plutonium from hydrogen bomb tests appeared in geological cores taken from twelve sites around the world.[1] During my lifetime the human population increased from just over two billion to eight billion, while the global temperature rose 1.1 degree Celsius.

The last time the world experienced such a rapid rise in temperature was in the Miocene, fourteen to seventeen million years ago. The immediate cause is the same, a rise in atmospheric carbon dioxide, but the underlying reasons differ. In the Miocene, the increase was caused by tectonic degassing of carbon; today, it is largely the result of burning fossil fuels.[2] The current speed of warming and carbon dioxide increase is unprecedented in the fossil record, with one exception, and that is the asteroid impact and volcanic eruptions that caused the decline of the dinosaurs sixty-six million years ago.[3] Associated with these contemporary changes, there has been widespread degradation of the environment, accelerating loss of biodiversity, and a magnification of economic and welfare inequalities among peoples.

The common thread running through my journey to becoming a scientist is a fascination with biological and cultural diversity and how this leads to change. Growing up in Arnside introduced me to the rich biodiversity of life and to an early realization, from the fossils that I found, that life had not been static over time. Studying biology and genetics at the University of Edinburgh gave me a solid foundation that prepared me for a study of Darwin's finches in pristine Galápagos with my husband, Peter, and our collaborators. This research revealed ways in which interactions between genetic, environmental, and social factors lead to changes that can set a population on a trajectory toward becoming a new species. Throughout, I was guided by the understanding, fostered by my parents, that a life devoted to science requires critical thinking, following exceptions to your pet theory, respect for others, and strong ethical values.

My interest in interaction between genes and the environment led naturally to our own species, *Homo sapiens,* a species capable of language and conscious thought, with a capacity to conceptualize ideas and communicate them to others. Such attributes have spurred outstanding creativity in science, medicine, music, art, and architecture but have also been used to incite racism, xenophobia, and war and exacerbate competition for natural resources. The result has been wide-scale destruction of the environment, global pollution, decreased biodiversity, and increased inequalities between peoples.

All organisms from bacteria to elephants modify their environment as they grow and reproduce and if left unchecked will rapidly expand in population numbers. Humans alone can recognize the causes and consequences of unregulated growth and have the capacity to collaborate with each other and reverse the process. I explored these questions more deeply after retirement, when Peter and I had the opportunity to travel while giving seminars and taking part in international workshops and scientific meetings.

We were fortunate to meet Indigenous peoples in Australia, South America, and North America and learn how they had lived for millennia within the limits of their local resources.

Asian and European history shows periods of great creativity interspersed with periods of intellectual and cultural decline. In the past, trade between civilizations brought people with diverse backgrounds together, and when the collaboration was peaceful, a flowering of creativity followed. Whether or not these alliances were diplomatic or violent was often determined by the attitude of a powerful leader, who could communicate with and have authority over thousands.[4] Both continents had examples of a rapid switch in the values and principles of a nation, which occurred under a change of leader with a different set of values and beliefs. The question then becomes, how do societies acquire a benevolent leader instead of a malevolent one? If a leader exercises force by subjugation and coercion, little can be done after the event except for protest. But if rhetoric and the hypnotic ability of a powerful orator are the tools used to persuade people, then an antidote lies in an informed and educated public. The essential feature is that an education that inspires critical thinking and respect for all peoples is a safeguard against propaganda and deception.

Having had the opportunity to participate in the education of both young children and university students, I saw the power of education. A stimulating education that fosters an individual's strengths, cooperation, creativity, love of learning, respect for others, and ethical values gives a person confidence and knowledge to think critically and be less easily swayed by misinformation. Cooperation between and mutual tolerance of people of different cultures has been the common feature during periods of creativity.

There is hope today that the meeting of minds will solve or at least mitigate our climate crisis and growing inequalities between peoples. As I discussed earlier in reference to Kristine Bonnevie (chapter 24), the key to change is often considered to be predicated on the role of the oppressed asserting their rights. This seldom works, but when meetings include those with the power to overturn entrenched institutionalized inequalities, remarkably rapid changes can occur. Communication is the key to resolving misunderstanding and when successful will rapidly provide much needed help for all who are disadvantaged, be it through gender, race, poverty, or other circumstances. This, I hope, will be the outcome of the coming together of prepared minds at the United Nations climate-change conferences.

In 2015, the United Nations Conference of the Parties, held in Paris with 189 countries participating, ratified an agreement to keep global temperatures from rising beyond 1.5 to 2 degrees Celsius by transitioning to renewable resources such as wind and solar power within fifteen years.[5] In December 2022, 190 governments from around the world joined the United Nations Biodiversity Conference of Parties (COP15) in Montreal.[6] Their goal was to transform the relationship between humans and biodiversity in such a way that by 2050 there would be a global vision of living in harmony with nature. Their framework acknowledged the contributions of Indigenous peoples as knowledgeable partners at the forefront of conservation and sustainable use of the environment, despite being among the world's most disadvantaged, poorest, and vulnerable people due to systemic marginalization.[7]

The Montreal meeting ended with the adoption of the Kunming-Montreal Global Biodiversity Framework (GBF), which addresses biodiversity loss, restoring ecosystems, and protection of Indigenous peoples' rights.[8]

The agreement includes four major targets:

1) Halt biodiversity loss and put 30 percent of the planet under protection by 2030.
2) Phase out subsidies that harm biodiversity by $500 billion per year while encouraging renewable resources. According to the International Monetary Fund, in 2022 the world spent $7 trillion on fossil-fuel subsidies, a $2 trillion increase since 2020.[9]
3) Increase financial flow from developed to developing countries by $30 billion per year.
4) Require transnational companies and financial institutions to monitor, assess, and transparently disclose risks and impacts on biodiversity through their operations.

To these we could add Partha Dasgupta's and Simon Levin's call for addressing our present-day imbalance between our global demands and nature's supply through a radical change in how our institutions think and measure economic success. They argue that in order to protect our own prosperity and nature, we must manage our global public goods, such as oceans and rain forests, instead of continually drawing on nature's commodities.[10] In other words, the economy of societies should be aligned with the economy of nature.

If the two treaties are implemented and our economic decisions pay respect to nature, climate disaster will be avoided, and we will be on the way to halting the loss of biodiversity and reducing

FIGURE 20. Upper: With Peter at Princeton University on the steps down to Eno Hall. Middle: Nicola (left) and Thalia (right). Lower (left to right), four grandchildren: Thalia and daughter Olivia; Anjali, now Orange, with Peter; Devon, with a butterfly on his hand; Rajul, examining a shell at a beach.

inequalities among peoples. It is through collaboration and cooperation between peoples from all cultures that we have hope for a peaceful, sustainable, and more equitable world for everyone, while keeping our dwindling biodiversity capable of further natural change.

By far the most important part of my life has been the thrill of sharing ideas and experiences with my husband, Peter, and together raising our two daughters in Canada, Galápagos, and the United States. The bouncing of unconstrained spontaneous ideas off each other daily, while respecting our many differences without reservation, not only led to more rigorous scientific research but was throughout my life an inspiration—whether when exploring nature, in an art gallery, listening to music, sitting with a glass of red wine by a log fire, or having a picnic by a lake. Going through life, in all its complexities, with such a companion, I treasure above all else.

My greatest wish is that in writing this book I have transmitted to others some of my own sense of wonder, joy, and excitement, in curiosity-driven research and in life, as they embark on their own journey. May they, too, find that touch of magic.

ACKNOWLEDGMENTS

I am exceedingly grateful to Sean McMahon, who encouraged me to write this book with the gift to both my husband, Peter, and myself of George Saunders's book *A Swim in a Pond in the Rain*. My warmest thanks go to Uli Reyer, Anne McClintock, Mark Chapin, Jessica Metcalf, my daughters, Nicola Chikkalingaiah and Thalia Grant, and three anonymous reviewers who read the first draft of the manuscript. All their many suggestions, comments, and advice were immensely helpful and seriously considered, and I adopted almost all of them. Peter and Nicola read and reread the manuscript many times, corrected errors, and provided valuable insights. To them I am enormously indebted. I have been extremely fortunate to have worked with an outstanding team at Princeton University Press. I mention especially Alison Kalett, for her wise and gentle editorial advice and encouragement, together with Hallie Schaeffer; Dimitri Karetnikov, who gave me much help with illustrations; Jill Harris, senior production editor, who guided me through the final stages; and Amy K. Hughes, an insightful, thoughtful copyeditor and a joy to work with. My greatest admiration and warmest thanks go to Peter, my closest companion and dearest friend for sixty-four years and counting.

APPENDIX A

We Are Not All the Same

All three of us Matchett siblings embarked on different career trajectories. Neither of my brothers was interested in biology. John developed his own highly successful business in computer software and was interested in flying and photography. His hallmark characteristics were a wholehearted commitment to a particular current interest, courage to tackle big problems, and independence. He had many interests and pursuits, apart from his business. He became a skilled helicopter pilot and was manager of and championship competitor for the British Helicopter Team, as well as a successful photographer. His first attempts at photography were playful tweaking of the photographs he took with an old bellows camera he found in the attic when he was a child, but he rapidly became more proficient and throughout his life won numerous prizes and accolades. One of his photographs, taken from the air, captures the billowing sail-like roof of the Sydney Opera House at dawn; another beautifully illustrates the sharply chiseled features of the giant sand dunes in Namibia; yet others depict the vitality of life amidst poverty in Cuba and shafts of light penetrating the clouds, reminiscent of an El Greco painting.

At school, John daydreamed through classes, much to my father's undisguised disgust. At age thirteen, he decided he wanted to play the organ. He was fascinated by the mechanics of it and the way in which the stops and pedals produce sounds through a variety of different-size pipes. He learned more by ear than by written music, and only one year later he was playing daily at the school's morning prayer service. During holidays, the Arnside Church committee asked him to play on Sundays. Not only had they lost their organist but the organ had lost two vital notes (B and D). I was amazed at his ability to rapidly change key to avoid having to play these delinquent notes. To John's delight, he was allowed to practice each day, but it turned out these were not just hymns and marches—instead, he spent long impious hours extemporizing and composing. When the church was almost visibly swaying to the loud jazzy music emanating from it, the church authorities threw him out. Silverdale, the neighboring village, snapped him up. Its organ had all the notes. It was well worth the five-mile bicycle ride, he thought, and most important, there were no restrictions on his creativity—the Silverdalers enjoyed it.

Elements of Andrew's characteristics as an adult were there as a child, especially his kindness, attention to detail, and aura of slight bewilderment. Our parents, always keen to encourage any interest expressed by their children, gave Andrew a kit to build a model of a James Watt steam engine for his eighth birthday. He worked on this meticulously for hours, days, and weeks, until one day, it was completed and ready to work. To get it going, he first tried lighting bits of twigs under the boiler, then a candle, but nothing happened. In typical Andrew fashion, he sat down, stared into space, and thought. Then, without saying anything, he got up, took money from his cashbox, and went down to the chemist. Half an hour later, he returned with a handful of small blue cubes of solid methylated spirits, which he stuffed into the space beneath the boiler. He lit one. The piston started slowly at first, then the wheels turned faster and then faster. Steam came out of the funnel. It even sounded like a steam engine! Andrew's eyes became wider and wider, the wheels went faster and faster, and the flames leapt higher and higher. Just as the expression on Andrew's face was changing from one of exhilaration to the slow realization of the possible consequences, my mother appeared as if from nowhere, running, with a wet blanket in her arms. As the engine became a hissing, heaving, smoldering lump under the blanket, and she was about to start on her "What ARE you children doing" speech, Andrew looked up with big blue eyes and said, "Mummy, it WORKED!" Even she had to smile.

Andrew followed exactly in my father's footsteps by studying medicine at the University of Glasgow and taking over my father's medical practice in Arnside. His other passions were painting in watercolors, sailing in a small wooden boat that he was constantly renovating, and woodwork, one creation being a beautiful quarter-size violin.

APPENDIX B

Nicola's Letter to Joel Achenbach, 2014

Hi Joel,

I would be very happy to talk about my parents and life on the Galapagos. I apologize for being slow in replying to you—this has been an unusually busy week, and I am actually writing this on a plane to California. I am the older of the two girls and went to the Galapagos every year with my parents and sister from the ages of 8 to 18. Quite simply, it was magical. I don't know if you have been to the Galapagos Islands, but for me they are like what the Celts call "thin places"—places where the veil between heaven and earth is frayed. My sister and I were very lucky to be able to spend a few months each year there. At one point we were there for the entire school year while my father was on sabbatical, and my mother homeschooled us. I don't remember ever being bored. We spent our days exploring whatever island we were on, swimming, inventing games, reading, and the older we got, the more we helped our parents with their research work. During the school year that we spent on the islands, my mother would teach us during the hottest part of the day, and the rest of the time we had to ourselves. Camping was rudimentary—we had no electricity and had to carefully conserve all our fresh water; they are "desert" islands after all. We had a few batteries for flashlights and a Walkman cassette player that my sister and I would share to listen to music in the evening. We rose and slept with the sun. We had to limit our luggage to one bag a piece so there was not a lot that we could bring. I read *The Lord of the Rings* one summer on Genovesa, an island in the north of the archipelago, and *War and Peace* on Daphne Major another year. We brought a violin and played it to the blue footed boobies. On Genovesa there was a saltwater lagoon that filled and emptied with the tides. A mother sealion used to leave her cub there while she went fishing, and we became friends.

I have always enjoyed teasing my parents mercilessly and I am pretty good at it. I claim that it keeps them grounded, but really it is just fun, and they are easy targets. There is one story that I use to tease my father which I will share. I was probably 13 or 14 years old, and we were hiking across a large expanse of unstable plates of spiky lava, the type they call a'a in Hawaii. We were all wearing hiking boots, shorts, tee shirts and had binoculars around our necks and notebooks in hand. We were in search of finches. My father usually sets the pace, and he has a long stride, so we were going at a fair clip. I mis-stepped and landed face forward flat on a nasty piece of lava. My parents swung round, my mother shouting "darling!" and my father shouting "Are the binoculars, ok?"

On a more heartfelt note, I would like to add a word about what it was like growing up as the daughter of my parents in general. My parents have always been passionate about their work, and it is always a joy to be around people who love what they do, so you can imagine how lucky we were to have parents who love their work. This could, of course, have gone either way, and my sister and I could have ended up playing second fiddle to a bunch of finches. What not many people know, however, is that they were equally passionate about being wonderful parents. They threw themselves into creating a loving, creative, adventurous home for my sister and me. We did have to train them a little and put limits on their discussions about their research, however, in order to keep them firmly focused on us. One night, at the dinner table, either my sister or I, I don't remember who, announced "NO TALKING SHOP AT THE TABLE." This was followed by complete silence as my parents sat, shocked, trying to quickly come up with something else to talk about.

APPENDIX C

Honors and Awards, with Some Comments

Peter and I have been honored many times with medals and prizes. The two of us have shared almost all of them. These are the most precious, since long-standing husband-and-wife research teams are a rarity. It is our hope they will become less rare.

I am very conscious of the fact that medals and prizes are rewards for research accomplishments, and that the recipients are not the only ones who performed the research. In our case, we were helped by graduate students and assistants in fieldwork and analysis and by colleagues working in laboratories with material that we supplied. Their help was indispensable.

No hierarchical ranking is implied by my comments below.

Medals and Awards. Peter and I received the Leidy Award from the Academy of Natural Sciences in Philadelphia (1994); the Darwin Medal (2002) and the Royal Medal (2017) from the Royal Society of London; and, with several other scientists, the Darwin-Wallace Medal from the Linnean Society (2008). These awards are humbling because they placed us in a long tradition of science carried out by eminent scientists. The E. O. Wilson Distinguished Naturalist Award of the American Society of Naturalists (1998), like some of the other prizes, carried the name of a scientist who influenced our research, yet it differed from the others in being the first of a newly created award program. The International Balzan Prize, which we received in 2005, aims "to promote culture, the sciences, and the most meritorious initiatives in the cause of humanity, peace, and fraternity among peoples throughout the world. The prizewinners must destine half of their awards for research projects carried out preferably by young humanists and scientists." With the money, we were able to support four young scientists and in addition hold a two-day international symposium in 2008, the year we officially retired. We invited top scientists in our broad area of interest, students, and collaborators, including four Ecuadorian students and their two professors. I mention the Kyoto Prize (2009) at length in chapter 23. In the words of Kazuo Inamori, it is an international award intended "to honor those who have contributed significantly to the scientific, cultural, and spiritual betterment of mankind." This echoes my parents' philosophy, and the award ceremony was a particularly moving experience for me, as it involved a week of interacting with people of all ages, from schoolchildren to university professors.

The BBVA Foundation Frontiers of Knowledge Award, presented to us in 2018, recognizes basic knowledge and the importance of the interdisciplinary nature of knowledge. Peter and I were in the company of James Allison, awarded for his immunological research that led to a successful cancer treatment; William Nordhaus, an economist and modeler of climate policy interventions such as carbon taxes; and Kaija Saariaho, a Finnish composer of contemporary music, selected for her excellence in music that, in the words of the foundation, "does much to shape the culture and sensibility of each era." My most recent award was the Frink Medal of the Zoological Society of London (2020), a medal bestowed for "significant and original contribution by a professional zoologist in the development of zoology." This was again humbling because of the list of eminent scientists who preceded me.

Named Awards. It is inspiring to receive an award in the name of a major figure in evolutionary biology and humbling when one's own name is attached to an award. In 2008, the Society for the Study of Evolution created the Rosemary Grant Advanced Award, a monetary prize for students at an advanced stage in their PhD research. I am extremely pleased that the beneficiaries are principally young, enthusiastic, and talented students at an important early stage of their careers. Having been inspired, I would like to inspire others. Three years later, the University of Zürich created an annual lecture named for both me and Peter.

Honorary Degrees. Peter and I have been jointly honored by McGill University (2000), Universidad San Francisco de Quito (2005), University of Zürich (2008), Ohio Wesleyan University (2012), University of Toronto (2017), and Princeton University (2019). Each of these honorary degrees is special in some way. The McGill degree gave me enormous pleasure because it was at McGill's library that I started to lay the foundation of my research, even though I did not have a position at the university. Furthermore, it was the first honorary degree Peter and I received jointly. Like the degrees from the University of Toronto and Princeton University, the honorary degree from the University of Zürich was the first to be jointly awarded in the institution's history. The Ohio Wesleyan degree was unique because neither of us had any connection with the university. The Toronto degree was unforgettable because it was awarded in Canada's 150th-anniversary year. The Princeton University degree means much to both of us because it was recognition from "our" university of thirty-four years. And the USFQ degree was special in being from the country we have worked in for more than four decades; moreover, the degree ceremony took place in Galápagos. In addition, I was also awarded an honorary PhD degree at the University of Helsinki in 2022. This one was special because the Finnish education system had been an inspiration for me throughout my teaching career.

NOTES

Chapter 1: Early Years

1. The Wellcome Collection's online record of my father's syringe: Science Museum Group, Inventor's prototype of cartridge syringe, A619671, Science Museum Group Collection Online, https://collection.sciencemuseumgroup.org.uk/objects/co141838/inventors-prototype-of-cartridge-syringe-hypodermic-syringes. The patent: Matchett, Alexander, Hypodermic injection appliances, United Kingdom, 805,031, nos. 31796/53 and 23848/54, class 81 (2), filed November 17, 1953, and August 17, 1954, issued December 15, 1954, https://patents.google.com/patent/GB805031A/en.

Chapter 2: War

1. B. Wicks, *No Time to Wave Goodbye* (Toronto: Stoddart, 1988). D. Prest, "Evacuees in World War Two—The True Story," BBC History, February 17, 2011, https://www.bbc.co.uk/history/british/britain_wwtwo/evacuees_01.shtml. Prest was the producer of BBC Radio 4's five-part series *Evacuation: The True Story* (1999), presented by Charles Wheeler. O. Sacks, *Uncle Tungsten: Memories of a Chemical Boyhood* (New York: Vintage, 2002).

Chapter 3: Living in a Medical Household

1. Excerpts from S. T. Coleridge, "The Rime of the Ancient Mariner" (1834 version), in *The New Oxford Book of English Verse, 1250–1950*, ed. H. Gardner (Oxford: Clarendon Press, 1972).

Chapter 6: Fossils in the Fire

1. I. R. Smith et al., "New Lateglacial Fauna and Early Mesolithic Human Remains from Northern England," *Journal of Quaternary Science* 28 (2013): 537–640.

2. F. Oldfield, "Studies in the Post-Glacial History of British Vegetation: Lowland Lonsdale," *New Phytologist* 59 (1960): 192–217.

3. A. Thom, *Megalithic Sites in Britain* (Oxford: Clarendon Press, 1971).

4. J. Rebanks, *The Shepherd's Life* (New York: Flatiron Books, 2015).

Chapter 7: School

1. B. Isacks, J. Oliver, and L. R. Sykes, "Seismology and the New Global Tectonics," *Journal of Geophysical Research* 73, no. 18 (1968): 5855, https://doi.org/10.1029/JB073i018p05855.

2. H. Harrer, *Seven Years in Tibet* (London: Rupert Hart-Davis, 1953).

3. C. Austen, *Adventures with Rosalind* (London: Hutchinson's Books for Young People, 1947).

4. C. Darwin, *On the Origin of Species by Means of Natural Selection, or Preservation of Favoured Races in the Struggle for Life* (London: John Murray, 1859).

5. Saint George's School, https://www.stge.org.uk.

6. J. N. Burstyn, "Education and Sex: The Medical Case Against Higher Education for Women in England, 1870–1900," *Proceedings of the American Philosophical Society* 117 (1973): 79–89.

Chapter 8: University

1. R. Buchsbaum, *Animals Without Backbones* (Chicago: University of Chicago Press, 1938); C. Auerbach, *Notes for Introductory Courses in Genetics* (Edinburgh: Oliver and Boyd, 1965).

2. D. S. Falconer, *Quantitative Genetics* (London: Longman, Harlow, 1960).

3. C. H. Waddington, "Genetic Assimilation of an Acquired Character," *Evolution* 7 (1953): 118–26.

4. G. H. Beale, *The Genetics of Paramecium aurelia* (Cambridge: Cambridge University Press, 1954).

5. B. R. Matchett, "A Study of the Immobilizing Antigens of *Acanthamoeba* sp." (honors BSc thesis, University of Edinburgh, 1960); held in American Philosophical Society Archives.

Chapter 9: Vancouver

1. P. R. Grant, *Enchanted by Daphne: The Life of an Evolutionary Naturalist* (Princeton: Princeton University Press, 2023).

2. J. B. Foster, "Evolution of Mammals on Islands," *Nature* 202 (1964): 234–35.

3. G. McMaster, *Iljuwas Bill Reid: Life and Work* (Toronto: Art Canada Institute, 2022).

4. C. H. Waddington, *Principles of Embryology* (London: George Allen and Unwin, 1956).

Chapter 10: Mexico

1. R. A. Clement et al., "Phylogeny, Migration and Geographic Range Size Evolution of *Anax* Dragonflies (Anisoptera: Aeshnidae)," *Zoological Journal of the Linnean Society* 194, no. 3 (2022): 858–78; M. Wikelski et al., "Simple Rules Guide Dragonfly Migration," *Biology Letters* 2, no. 3 (2006): 325–29.

Chapter 11: Marriage

1. P. R. Grant, "The Adaptive Significance of Some Size Trends in Island Birds," *Evolution* 19 (1965): 355–67.

2. M. W. Nirenberg and H. J. Matthaei, "The Dependence of Cell-Free Protein Synthesis upon Naturally Occurring or Synthetic Polyribonucleotides," *Proceedings of the National Academy of Sciences USA* 47 (1961): 1588–1602.

Chapter 12: Rosemary's Monday

1. E. Arnason and P. R. Grant, "Climatic Selection in *Cepaea hortensis* at the Northern Limit of Its Range in Iceland," *Evolution* 30 (1976): 499–508.

2. P. R. Grant, "Interactive Behaviour of Puffins (*Fratercula arctica* L.) and Skuas (*Stercorarius parasiticus* L.)," *Behaviour* 40 (1971): 262–81.

3. E. Jarvis, "Evolution of Vocal Learning and Spoken Language," *Science* 366, no. 6461 (2019): 50–54.

4. B. Magnússon et al., "Seabirds and Seals as Drivers of Plant Succession on Surtsey," *Surtsey Research* 14 (2020): 115–30.

5. H. Ingstad and A. S. Ingstad, *The Viking Discovery of America: The Excavation of a Norse Settlement in L'Anse Aux Meadows, Newfoundland* (New York: Checkmark Books, 2001).

6. M. Price, "Marking Time," *Science* 380, no. 6641 (2023): 124–28.

7. N. Calver and M. Parker, "The Logic of Scientific Unity? Medawar, the Royal Society and the Rothschild Controversy 1971–72," *Notes and Records: The Royal Society Journal of the History of Science* 70, no. 1 (2016).

8. M. Curie, *The Discovery of Radium: Address by Madame M. Curie at Vassar College, May 14, 1921*, Ellen S. Richards Monographs No. 2 (Poughkeepsie, NY: Vassar College, 1921).

Chapter 13: Thwarted

1. P. R. Grant, "Convergent and Divergent Character Displacement," *Biological Journal of the Linnean Society* 4 (1972): 39–68.

Chapter 14: One Step Sideways

1. A. Patchett, *Bel Canto* (New York: Perennial, HarperCollins, 2001).

2. P. Sahlberg, *Finnish Lessons 3.0: What Can the World Learn from Educational Change in Finland?*, 3rd ed. (New York: Teachers College Press, 2021).

3. Kerola Marika, quoted in E. Benke and M. Spring, "US Midterm Elections: Does Finland Have the Answer to Fake News?," *BBC News*, October 12, 2022, https://www.bbc.com/news/world-europe-63222819.

Chapter 15: Three Steps Forward

1. C. Darwin, *On the Origin of Species by Means of Natural Selection, or Preservation of Favoured Races in the Struggle for Life* (London: John Murray, 1859).

2. A. R. Wallace, Letter to H. W. Bates, October 11, 1847, in P. Raby, *Alfred Russel Wallace: A Life* (Princeton, NJ: Princeton University Press, 2001).

3. S. J. Wagstaff et al., "Classification, Origin, and Diversification of the New Zealand *Hebes* (Scrophulariaceae)," *Annals of the Missouri Botanical Garden* 89, no. 1 (2002): 38–63.

4. D. Lack, *Darwin's Finches* (Cambridge: Cambridge University Press, 1947).

5. A. Caccone et al., "Phylogeography and History of Giant Galápagos Tortoises," *Evolution* 56 (2002): 2052–66.

6. F. Orellana-Rovirosa and M. Richards, "Emergence/Subsidence Histories along the Carnegie and Cocos Ridges and Their Bearing upon Biological Speciation in the Galápagos," *Geochemistry, Geophysics, Geosystems* 19 (2018), 4099–4129.

7. A. Cox and G. Dalrymple, "Palaeomagnetism and Potassium-Argon Ages of Some Volcanic Rocks from the Galápagos Islands," *Nature* 209 (1966), 776–77.

8. B. R. Grant and P. R. Grant, *Evolutionary Dynamics of a Natural Population: The Large Cactus Finch of the Galápagos* (Chicago: Chicago University Press, 1989).

Chapter 16: Daphne Research

1. P. R. Grant and B. R. Grant, *40 Years of Evolution: Darwin's Finches on Daphne Major Island* (Princeton, NJ: Princeton University Press, 2014).

2. D. Lack, *Darwin's Finches* (Cambridge: Cambridge University Press, 1947); P. R. Grant, *Ecology and Evolution of Darwin's Finches* (Princeton, NJ: Princeton University Press, 1986).

3. S. Lamichhaney et al., "Evolution of Darwin's Finches and Their Beaks Revealed by Genome Sequencing," *Nature* 518 (2015): 371–75.

4. S. Lamichhaney et al., "A Beak Size Locus in Darwin's Finches Facilitated Character Displacement during a Drought," *Science* 352 (2016): 470–74.

5. P. R. Grant and B. R. Grant, "Evolution of Character Displacement in Darwin's Finches," *Science* 313 (2006): 224–26; Lamichhaney et al., "A Beak Size Locus."

6. L. Andersson, "Genes Controlling Beak Size and Shape in Darwin's Finches," *Access Science* (McGraw-Hill Education), 2017; S. Lamichhay et al., "Female-Biased Gene Flow Between Two Species of Darwin's Finches," *Nature Ecology and Evolution* 4 (2020): 979–86.

7. Lamichhaney et al., "Female-Biased Gene Flow."

8. P. R. Grant and B. R. Grant, "Triad Hybridization via a Conduit Species," *Proceedings of the National Academy of Sciences USA* 117 (2020): 7888–96.

9. E. D. Enbody et al., "Community-Wide Genome Sequencing Reveals 30 Years of Darwin's Finch Evolution," *Science* 381 (2023), eadf6218.

10. S. Lamichhaney et al., "Rapid Hybrid Speciation in Darwin's Finches," *Science* 359 (2018): 224–28.

11. S. Pääbo, "The Diverse Origins of the Human Gene Pool," *National Review of Genetics* 16 (2015), 313–14.

12. H. Zeberg and S. Pääbo, "The Major Genetic Risk for Severe COVID-19 Is Inherited from Neanderthals," *Nature* 587 (2020): 610–12; S. Zhou et al., "A Neanderthal *OAS1* Isoform Protects Individuals of European Ancestry Against COVID-19 Susceptibility and Severity," *Nature Medicine* 27, no. 4 (2021): 659–67; H. Zeberg and S. Pääbo, "A Genomic Region Associated with Protection Against Severe COVID-19 Is Inherited from Neandertals," *Proceedings of the National Academy of Sciences USA* 118 (2021), e2026309118.

13. R. Xie et al., "The Episodic Resurgence of Highly Pathogenic Avian Influenza H5 Virus," *Nature* 622 (2023): 810–17.

14. M. Greaves and C. Maley, "Clonal Evolution in Cancer," *Nature* 481 (2012): 306–13.

15. K. H. Burns, "Transposable Elements in Cancer," *Nature Reviews Cancer* 17 (2017): 415–24.

16. K. H. Kapralova et al., "Evolution of Adaptive Diversity and Genetic Connectivity in Arctic Charr (*Salvelinus alpinus*) in Iceland," *Heredity* 106, no. 3 (2011), 472–87.

17. M. Brachmann et al., "Variation in the Genomic Basis of Parallel Phenotypic and Ecological Divergence in Benthic and Pelagic Morphs of Icelandic Arctic Charr (*Salvelinus alpinus*)," *Molecular Ecology* 31 (2022): 4688–4706; L. Andersson and M. Purugganan, "Molecular Genetic Variation of Animals and Plants under Domestication," *Proceedings of the National Academy of Sciences USA* 119, no. 30 (2022): p.e2122150119.

Chapter 17: The Magical Years as a Family on Genovesa and Daphne

1. P. R. Grant and N. Grant, "Breeding and Feeding of Galápagos Mockingbirds, *Nesomimus parvulus*," *Auk* 96 (1979): 723–36.

2. P. R. Grant and K. T. Grant, "Breeding and Feeding Ecology of the Galápagos Dove," *Condor* 81 (1979): 397–403.

Chapter 18: Teaching and Research in Princeton

1. C. Wedekind and S. Fúri, "Body Odour Preferences in Men and Women: Do They Aim for Specific MHC Combinations or Simply Heterozygosity?," *Proceedings of the Royal Society B*, 264 (1997): 1471–79.

Chapter 19: Interlude in Nepal

1. S. McLennan, "Music Preview: Tabla Superstar Zakir Hussain—Do Not Fear a Sonic Culture Clash," *Arts Fuse*, March 21, 2016, https://artsfuse.org/142552/fuse-music-preview-tabla-superstar-zakir-hussain-do-not-fear-a-sonic-culture-clash/.

2. G. R. Scott et al., "How Bar-headed Geese Fly over the Himalayas," *Physiology (Bethesda)* 2 (2015): 107–15.

3. G. Modiano et al., "Protection Against Malaria Morbidity: Near-Fixation of the α-Thalassemia Gene in a Nepalese Population," *American Journal of Human Genetics* 48 (1991): 390–97.

4. P. Zhang et al., "Denisovans and *Homo sapiens* on the Tibetan Plateau: Dispersal and Adaptations," *Trends in Ecology and Evolution* 37 (2022): 257–67.

Chapter 20: Retirement

1. J.N.M. Smith et al., *Conservation and Biology of Small Populations: The Song Sparrows of Mandarte Island* (Oxford: Oxford University Press, 2006).

2. C. Valle, "Ecological Selection and Evolution of Body Size and Sexual Dimorphism in the Galápagos Flightless Cormorant," in *Evolution from the Galapagos: Social and Ecological Interactions in the Galapagos Islands*, vol. 2, ed. G. Trueba and C. Montúfar (New York: Springer, 2013).

3. T. Brook, *Vermeer's Hat* (New York: Bloomsbury Press, 2009).

Chapter 21: Indigenous Peoples

1. D. S. London and B. Beezhold, "A Phytochemical-Rich Diet May Explain the Absence of Age-Related Decline in Visual Acuity of Amazonian Hunter-Gatherers in Ecuador," *Nutrition Research* 35 (2015): 107–17.

2. C. Vizcarra et al., "Spatial Distribution of Oil Spills in the Northeastern Ecuadorian Amazon: A Comprehensive Review of Possible Threats," *Biological Conservation* 252 (2020): 108820.

3. J. G. Blake and B. A. Loiselle, "Enigmatic Declines in Bird Numbers in Lowland Forest of Eastern Ecuador May Be a Consequence of Climate Change," *PeerJ* 3 (2015): e1177; J. G. Blake and B. A. Loiselle, "Long-Term Changes in Composition of Bird Communities at an 'Undisturbed' Site in Eastern Ecuador," *Wilson Journal of Ornithology* 128, no. 2 (2016): 255–67.

4. D. L. Wagner et al., "Insect Decline in the Anthropocene: Death by a Thousand Cuts," *Proceedings of the National Academy of Sciences USA* 118, no. 2 (2021), e2023989118.

5. H. Ralimanana et al., "Madagascar's Extraordinary Biodiversity: Threats and Opportunities," *Science* 378 (2022): 963.

6. Jonathan Watts, "Witness to Paradise Being Lost: My Year in the Dying Amazon," *Guardian*, December 16, 2022, https://www.theguardian.com/environment/2022/dec/16/year-in-the-life-of-the-amazon-deforestation-climate-disaster-mass-extermination.

Chapter 22: Australia

1. S. K. May et al., "New Insights into the Rock Art of Anbangbang Gallery, Kakadu National Park," *Journal of Field Archaeology* 45, no. 2 (2020): 120–34.

2. B. Phillips and R. Shine, "Adapting to an Invasive Species: Toxic Cane Toads Induce Morphological Change in Australian Snakes," *Proceedings of the National Academy of Sciences USA* 101 (2004), 17150–55.

Chapter 23: The Diversity of Asia

1. Sima Qian and K. E. Brashier, *The First Emperor: Selections from the Historical Records* (New York: Oxford University Press, 2007).

2. D. Quinbo, "Sino-Western Cultural Exchange as Seen Through the Archaeology of the First Emperor's Necropolis," *Journal of Chinese History* 7, no. 1 (2023): 21–72.

3. L. L. Cavalli-Sforza and F. Cavalli-Sforza, *The Great Human Diasporas* (New York: Addison-Wesley, 1995).

4. H. Harrer, *Seven Years in Tibet* (London: Rupert Hart-Davis, 1953).

5. Vincent van Gogh, *The Potato Eaters*, https://www.vangoghmuseum.nl/en/collection/s0005V1962.

Chapter 24: Return to Europe

1. R. Holmes, *The Age of Wonder* (New York: Vintage Books, 2010), chap. 4.

2. I. Hanski, *Messages from Islands: A Global Biodiversity Tour* (Chicago: Chicago University Press, 2016).

3. C. Posth et al., "Palaeogenomics of Upper Palaeolithic to Neolithic European Hunter-Gatherers," *Nature* 615 (2023): 117–26.

4. M. Kohler and S. Moya-Sola, "Physiological and Life History Strategies of a Fossil Large Mammal in a Resource-Limited Environment," *Proceedings of the National Academy of Sciences USA* 106, no. 48 (2009): 20354–58.

5. E. Coen, *Cells to Civilizations: The Principles of Change That Shape Life* (Princeton, NJ: Princeton University Press, 2012).

6. R. Cámara-Leret and J. Bascompte, "Language Extinction Triggers the Loss of Unique Medicinal Knowledge," *Proceedings of the National Academy of Sciences USA* 118, no. 24 (2021), e2103683118.

7. R. Ross Holloway, "The Tomb of the Diver," *American Journal of Archaeology* 110, no. 3 (2006): 365–88.

Chapter 25: Where Do We Go from Here?

1. C. N. Waters et al., "Defining the Onset of the Anthropocene," *Science* 378 (2022): 706–8.

2. T. D. Herbert et al., "Tectonic Degassing Drove Global Temperature Trends since 20 Ma," *Science* 377 (2022): 116–19.

3. C. H. Lear et al., "Geological Society of London Scientific Statement: What the Geological Record Tells Us about Our Present and Future Climate," *Journal of the Geological Society* 178, no. 1 (2021): 220–39.

4. L. L. Cavalli-Sforza and F. Cavalli-Sforza, *The Great Human Diasporas* (New York: Addison-Wesley, 1995).

5. M. Masaquiza Jerez, "Challenges and Opportunities for Indigenous Peoples' Sustainability," April 23, 2021, United Nations Department of Economic and Social

Affairs: Social Inclusion, https://www.un.org/development/desa/dspd/2021/04/indigenous-peoples-sustainability/.

6. Convention on Biological Diversity, "COP 15: Nations Adopt Four Goals, 23 Targets for 2030 in Landmark UN Biodiversity Agreement," press release, December 19, 2022, https://www.cbd.int/article/cop15-cbd-press-release-final-19dec2022; "COP15 Ends with Landmark Biodiversity Agreement," December 20, 2022, UN Environment Programme, https://www.unep.org/news-and-stories/story/cop15-ends-landmark-biodiversity-agreement.

7. M. Halewood, "Reporting from COP15: Digital Sequence Information Will Influence Who Benefits from Biodiversity," blog post, December 19, 2022, Alliance, Biodiversity & CIAT, https://alliancebioversityciat.org/stories/cop15-digital-sequence-information.

8. Convention on Biological Diversity (CBD), "Kunming-Montreal Global Biodiversity Framework," report, December 19, 2022, UN Environment Programme, https://www.unep.org/resources/kunming-montreal-global-biodiversity-framework.

9. "Climate Change: Fossil Fuel Subsidies," International Monetary Fund, https://www.imf.org/en/Topics/climate-change/energy-subsidies.

10. P. Dasgupta and S. Levin, "Economic Factors Underlying Biodiversity Loss," *Philosophical Transactions of the Royal Society B* 378 (2023), 20220197.

INDEX

A NOTE ON THE TYPE

This book has been composed in Arno, an Old-style serif typeface in the classic Venetian tradition, designed by Robert Slimbach at Adobe.